Forest Governance and Management Across Time

The influence of the past, and of the future on current-time trade-offs in the forest arena are particularly relevant given the long-term successions in forest landscapes and the hundred years' rotations in forestry. Historically established path dependencies and conflicts determine our present situation and delimit what is possible to achieve. Similarly, future trends and desires have a large influence on decision-making. Nevertheless, decisions about forest governance and management are always made in the present – in the present-time appraisal of the developed situation, future alternatives and in negotiation between different perspectives, interests, and actors.

This book explores historic and future outlooks as well as current trade-offs and methods in forest governance and management. It emphasizes the generality and complexity with empirical data from Sweden and inter-nationally. It first investigates, from a historical perspective, how previous forest policies and discourses have influenced current forest governance and management. Second, it considers methods to explore alternative forest futures and how the results from such investigations may influence the present. Third, it examines current methods of balancing trade-offs in decision-making among ecosystem services. Based on the findings the authors develop an integrated approach – Reflexive Forestry – to support exchange of knowledge and understandings to enable capacity building and the establishment of common ground. Such societal agreements, or what the authors elaborate as forest social contracts, are sets of relational commitment between involved actors that may generate mutual action and a common directionality to meet contemporary challenges.

Erland Mårald is Professor in History of Science and Ideas at Umeå University, Sweden.

Camilla Sandström is Professor in Political Science at Umeå University, Sweden, and the deputy Program Director of the Future Forests research programme.

Annika Nordin is Professor in Forest Ecophysiology at the Swedish University of Agricultural Sciences and the Program Director of the Future Forests research programme.

The Earthscan Forest Library

This series brings together a wide collection of volumes addressing diverse aspects of forests and forestry and draws on a range of disciplinary perspectives. Titles cover the full range of forest science and include the biology, ecology, biodiversity, restoration, management (including silviculture and timber production), geography and environment (including climate change), socio-economics, anthropology, policy, law and governance. The series aims to demonstrate the important role of forests in nature, peoples' livelihoods and fin contributing to broader sustainable development goals. It is aimed at undergraduate and postgraduate students, researchers, professionals, policy-makers and concerned members of civil society.

Series Editorial Advisers:

John L. Innes, Professor and Dean, Faculty of Forestry, University of British Columbia, Canada.

John Parrotta, Research Program Leader for International Science Issues, US Forest Service – Research & Development, Arlington, Virginia, USA.

Jeffrey Sayer, Professor and Director, Development Practice Programme, School of Earth and Environmental Sciences, James Cook University, Australia, and Member, Independent Science and Partnership Council, CGIAR (Consultative Group on International Agricultural Research).

Recent Titles:

Forest Governance and Management Across Time
Developing a New Forest Social Contract
Edited by Erland Mårald, Camilla Sandström and Annika Nordin and others

Sustainable Forest Management
From Concept to Practice
Edited by John L. Innes and Anna V. Tikina

Gender and Forests
Climate Change, Tenure, Value Chains and Emerging Issues
Edited by Carol J. Pierce Colfer, Bimbika Sijapati Basnett and Marlène Elias

Forests, Business and Sustainability
Edited by Rajat Panwar, Robert Kozak and Eric Hansen

Climate Change Impacts on Tropical Forests in Central America
An Ecosystem Service Perspective
Edited by Aline Chiabai

Rainforest Tourism, Conservation and Management
Challenges for Sustainable Development
Edited by Bruce Prideaux

Large-scale Forest Restoration
David Lamb

Additional information on these and further titles can be found at
www.routledge.com/books/series/ECTEFL

Forest Governance and Management Across Time

Developing a New Forest Social Contract

Main authors
Erland Mårald, Camilla Sandström and Annika Nordin

Contributing authors
Lucy Rist, Anna Sténs, Karin Beland Lindahl,
Annika Carlsson-Kanyama, Johanna Johansson,
Carina Keskitalo, Hjalmar Laudon, Rolf Lidskog,
Tomas Lämås, Tomas Lundmark, Urban Nilsson,
Eva-Maria Nordström, Jean-Michel Roberge
and Johan Sonesson

LONDON AND NEW YORK

First published 2017 by Routledge

2 Park Square, Milton Park, Abingdon, Oxfordshire OX14 4RN
52 Vanderbilt Avenue, New York, NY 10017

Routledge is an imprint of the Taylor & Francis Group, an informa business

First issued in paperback 2019

British Library Cataloguing-in-Publication Data
A catalogue record for this book is available from the British Library

Library of Congress Cataloging-in-Publication Data
A catalog record for this book has been requested

ISBN: 978-1-138-90430-9 (hbk)
ISBN: 978-0-367-35140-3 (pbk)

Typeset in Sabon
by Florence Production Ltd, Stoodleigh, Devon, UK

Contents

Figures

Tables

Main authors and contributors

Main authors

Erland Mårald is a professor in the history of science and ideas at the Department of Historical, Philosophical and Religious Studies, Umeå University. His research focuses on the history of agricultural and forest sciences, and environmental ideas.

Camilla Sandström is a professor in political science at the Department of Political Science, Umeå University. Her research interests include the governance and management of natural resources with a particular focus on institutional aspects and conflict management.

Annika Nordin is a professor in ecophysiology at the Department of Forest Genetics and Plant Physiology, Umeå Plant Science Centre, Swedish University of Agricultural Sciences. Her research interests include inter-disciplinary forest science and science to policy communication.

Contributors

Lucy Rist is a PhD and researcher at the Department of Forest Ecology and Management at the Swedish University of Agricultural Sciences. Her research interests include ideas about sustainability and natural resource management, both how sustainability is defined and how it may be reached. Much of her work is inter- or transdisciplinary.

Anna Sténs is a PhD and researcher in history at the Department of Historical, Philosophical and Religious Studies, Umeå University. Her research focuses on contemporary social, political and environmental history.

Karin Beland Lindahl is an assistant professor at the Political Science Unit, Luleå University of Technology. Her research interest is in politics of natural resource management, particularly interpretive policy analysis and conflict management.

Annika Carlsson Kanyama is reader and a researcher at KTH, The Royal Institute of Technology, Division of Industrial Ecology. Her research interests include methodologies for future studies, adaptation to climate change and environmental impacts of consumption patterns.

Johanna Johansson is a senior lecturer in environmental social sciences at Södertörn University. Her research interest is in the policy design and outcomes of Swedish forest policy, particularly the role played by deliberative processes.

E. Carina H. Keskitalo is a professor of political science at the Department of Geography and Economic History, Umeå University. She focuses on environmental policy in relation to societal and environmental change (among other climate change).

Tomas Lämås is an associate professor in forest management planning at the Department of Forest Resource Management, Swedish University of Agricultural Sciences. His research interest is in decision support systems for multi objective forestry.

Hjalmar Laudon is a professor of forest landscape biogeochemistry at the Department of Forest Ecology and Management, Swedish University of Agricultural Sciences. His research interests are primarily related to hydrology and biogeochemistry in the forested landscape and questions concerning the role of connectivity, scaling, forestry impact and climate change.

Rolf Lidskog is a professor at the Environmental Sociology Section, Örebro University, Sweden. His research focuses on environmental regulation, especially the role of expertise in environmental governance.

Tomas Lundmark is a professor of forest management at the Department of Forest Ecology and Management, Swedish University of Agricultural Sciences. His research focuses on the effects of management of boreal forests on carbon balance and climate change mitigation potential.

Urban Nilsson is a professor in forest production at the Southern Swedish Forest Research Centre, Swedish University of Agricultural Sciences. His research interests are silviculture in monocultures of conifer species as well as in mixed species stands.

Eva-Maria Nordström is an associate professor of forest resource management at the Department of Forest Resource Management, Swedish University of Agricultural Sciences. Her research interests include forest management planning and scenario analysis, with a focus on situations with conflicting objectives and multiple stakeholders.

Jean-Michel Roberge is an associate professor of forest ecology at the Department of Wildlife, Fish and Environmental Studies, Swedish University of Agricultural Sciences. His research interests include conservation biology and restoration ecology with special emphasis on boreal forest landscapes.

Johan Sonesson is a PhD and researcher at the Forestry Research Institute of Sweden. His research interests covers silviculture, forest management and planning.

Preface

Forest governance and management is conducted in the strait between the past and the future. Historically established path dependencies, institutions, discourses, legislation, and conflicts determine our present situation and delimit what is possible to achieve. Similarly, future trends, projections and desires have a large influence on current decision-making. The influence of the past, and of the future on current-time tradeoffs, are particularly relevant for forest governance and management, given the long-term successions in forest landscapes and the hundred years' rotations in forestry. Nevertheless, decisions about forest governance and management are always made in the present – in the present-time appraisal of the developed situation, future alternatives, and in negotiation between different perspectives, interests, and actors.

Forest governance and management likewise connect to specific socio-ecological circumstances. Research about globalization and diverging problem framings have recognized a much more complex society with an increasing array of actors and institutions at multiple levels. Consequently, it has become increasingly clear that research, governance, and management need to be conducted in interaction with stakeholders, for example, local actors, and citizens in different kinds of collaborative, social learning and pathway developing processes. Behind such approaches lies the assumption that issues that span between social and ecological spheres, and between different scientific disciplines, value systems and temporal dimensions cannot be solved in piecemeal and by experts only.

This book explores and assesses these historic and future outlooks as well as current tradeoffs and integrated approaches to reach sustainable development. It emphasizes the generality and complexity in forest governance and management, illustrating this with empirical data both Swedish and international. We first investigate, from a historical perspective, how previous forest policies and discourses have influenced current forest governance and management. Second, we explore methods to realize alternative forest futures and how the results from such investigations may influence the present. Third, we examine current methods of balancing tradeoffs in decision-making among ecosystem services, and how these methods relate to both what can be learnt from the past and what is expected from the future. Based on the findings we then develop an integrated approach – Reflexive Forestry – to support exchange of knowledge, practices and understandings to enable social capacity building and the establishment of common ground in the forest arena. Such societal agreements, or what we call *forest social contracts*, are sets of implicit and/or explicit arrangements built on trust and relational commitment between involved actors to achieve objectives. Hence, through competing claims and understandings, and through mutual cooperation, compromise and commitments by stakeholders, a common directionality may be generated and sustained to meet contemporary challenges.

This book has been produced within the interdisciplinary research programme Future Forests. It was conducted during the period 2009–2016, hosted by the Swedish University of Agricultural Sciences (SLU) in collaboration with Umeå University and the Forestry Research Institute of Sweden (Skogforsk), with the aim of addressing complex problems relating to forest use in the context of climate change. In total, the programme has generated c. 350 peer-reviewed articles, mainly within disciplines in the natural sciences, social sciences and the humanities, but also through a substantial number of interdisciplinary articles crossing these scientific fields. In the programme stakeholder involvement has been key, as has been an ambitious commitment to dialogue and communication.

The Future Forests programme has thus paved the way for this book. The book is, however, not an effort to summarize or synthesize the programme. Although many references to research findings from the programme, this problem driven book presents original research and aims to formulate general conclusions that will be highly interesting for international forest researchers, forest manager practitioners and forest policy-makers. It has been very challenging to write a coherent and argumentative book with so many authors involved from different scientific disciplines, including different epistemologies, research traditions and literary styles. In practice, the involved authors have contributed with text, and the main authors have, in communication with the wider author-collective, formulated the overarching structure and themes of the book, and compiled the different parts into a coherent text. Finally, all authors have read and commented on the manuscript, and agreed on the main arguments and the content of the book.

Acknowledgements

The idea to write a book within Future Forests started in Phase I of the programme (2009–2012), initiated by professor Stig Larsson and professor Jon Moen to whom we are greatly indebted. The idea was also supported by the Board of Future Forests and we want to thank the chair of the Board, Maria Norrfalk, who together with the Board members (during Phase II, 2013–2016) Wilhelm Agrell, Marianne Eriksson, Martin Holmberg, Olof Johansson, Lotta Möller, Håkan Wirtén, and Thomas Nilsson has been supporting us through the process. The book project has also been discussed several times with all researchers in Phase II of Future Forests and therefore we thank: Johan Bergh, Mats Berlin, Kevin Bishop, Christer Björkman, Johanna Boberg, David Ellison, Gustaf Egnell, Karin Eklöf, Kristina Espmark, Nils Fahlvik, Adam Felton, Nicklas Forsell, Martyn Futter, Peichen Gong, Anna Gunulf, Artti Juutinen, Maartje J. Klapwijk, Florian Kraxner, Hanna Lundmark, Lars Lundqvist, Anders Lundström, Erik Löfmarck, Tuomas Nummelin, Maria Pettersson, Thomas Ranius, Maria Riala, Eva Ring, Daniel Sjödin, Ryan Sponseller, Caroline Strömberg, Kristina Wallertz, Camilla Widmark, Anneli Ågren, and Lars Östlund.

During the work with the book we have regularly interacted with many different actors in various dialogue processes, workshops, excursions and collaborations, and we thank you all. Especially we want to thank those of you that have participated in our reference groups during Phase II: Göran Bergqvist, Jesper Runge, Jonas Bergqvist, Per Simonsson, Stina Moberg, Dan Glöde, Carl Agestam, Elisabet Andersson, Gunilla Forsgren Johansson, Per Olsson, Stefan Bleckert, Mats Blomberg, Lennart Henriksson, Fredrik Klang, Hillevi Eriksson, Åke Granqvist, Anna Lundborg, Karin Vestlund Ekerby, Björn Boström, Britta Wännström, Jonas Eriksson, Johanna Fintling, Göran Örlander, Jan Terstad, Göran Andersson, Viktoria Hallberg, Malin Andersson, Fredrik Widemo, Henrik von Stedingk, Linda Eriksson, and Maria Boström.

Earlier versions of the manuscript have been read and commented on by professor Katarina Eckerberg, professor Stig Larsson, and professor Lars Östlund, who provided valuable inputs and influenced the direction of the book project. We want to thank Dr John Blackwell/Sees-Editing for the editing of the final manuscript and Jerker Lokrantz/Azote for the design of the illustrations. Finally, we thank the staff in Future Forests – Annika Mossing, Linda Gruffman, and Jan-Peter Nordmark – for helping us solve many practical problems, and journalists Lars Klingström and Mats Hannertz who inspired us regarding communication. This book is a part of the Future Forests programme funded by Mistra (the Swedish Foundation for Strategic Environmental Research), the Swedish University of Agricultural Sciences (SLU), Umeå University, Skogforsk, and the Swedish forest industry.

Umeå 21 March 2017
Erland Mårald, Camilla Sandström
and Annika Nordin

Introduction

A major theme in contemporary debate concerns the optimal ways to promote changes that improve social and environmental conditions for both current and future generations. During the last three decades, global society has become increasingly goal-oriented. Several United Nations (UN) initiatives have contributed to this development, including the Brundtland Commission's establishment of the principle of sustainable development in 1987, the Earth Summit in Rio de Janeiro and Agenda 21 in 1992, the Millennium Development Goals in 2000, and the Sustainable Development Goals in 2015 (Clark and Dickson 2003; Lafferty and Eckerberg 2013; Griggs *et al.* 2013). Similarly, various specific international goals have been agreed, the most well-known perhaps being the target to limit global warming to below 2°C, or even 1.5°C, set by the Paris Agreement in 2015 (UNFCCC 2015). On a regional level, the European Union has identified various societal challenges and, for example, initiated the Europe 2030 sustainable agenda, which sets various specific goals (European Commission 2015). All these goals are future-oriented and intended to mobilize resources and promote processes that facilitate the implementation of solutions. These solutions focus on urgent problems, capitalizing on opportunities as they emerge, as well as stimulating transitions towards the stated goals. In this manner, the establishment of ambitious goals provides a way to show political leadership and foster desired change, or at least to signpost the direction of desired change.

This strong focus on goals has coincided with an international trend towards liberalization and deregulation of governance and markets (Ong 2006; Harvey 2007). This trend has affected socio-economic and political activities in virtually all sectors, including international and Swedish forest policy, management and arenas (the main foci of this book) from the early 1990s onwards. Leading approaches to implement forest policy are now: marketization, enhancement of the private sector's role, deregulation, and voluntarism (Humphreys 2006; 2014). The state has turned from a hierarchically organised regulator to a co-ordinator setting the major goals, releasing market forces, and activating incentives for competition and development (Appelstrand 2012). To monitor developments in this deregulated context, and to assess degrees to which goals have been fulfilled, accountability has become crucial (Hood 1995). Indeed, assessments and evaluations have become so extensive that it has been claimed that we now live in an 'audit society' (Power 1997). It could also be regarded as a 'goal society', since the establishment of goals and the audit explosion are inextricably linked.

These observations raise obvious questions about the effectiveness of this shift to governing by setting goals (which can be illustrated by the changes in numbers of goals for the governance and management of forests in Sweden between 1850 and 2015; Figure 1). In 1993 a new Forestry Act set out two main and 'equal' objectives – to provide a valuable yield and preserve biodiversity (Skogsvårdslag 1993). The Act also states that social, cultural, and aesthetic values, and Sami reindeer herding, should be considered. Furthermore, fifteen National Environmental Quality Objectives were established in 1999, and supplemented in 2005 with another objective concerning biodiversity (Sweden's Environmental Objectives Council 2006). Beside biodiversity, objectives regarding climate, water, wetlands, eutrophication, acidification and non-toxic environments were all relevant (to various degrees) to forestry. Thus, demands placed on forestry have expanded from a narrow requirement to maintain yields, to obligations or expectations regarding increasingly large numbers of goals.

Of course, simply increasing numbers of goals does not necessarily improve forest governance and management. In Sweden, the resulting 'goal inflation' has been accompanied by an optimistic view that has been called 'more-of-everything', a belief that more can be extracted from existing resources without resolving inherent goal conflicts (Beland Lindahl *et al.* 2015). However, implementation of the increasingly wide range of goals heavily relies on voluntary instruments with weak trade-off mechanisms,

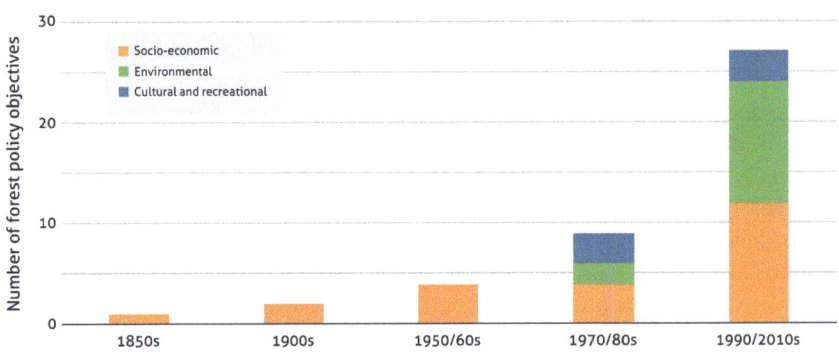

Figure 0.1 Changes in numbers of explicitly mentioned 'goals' or main directions in Swedish legislation from 1850 to 2015.

Sources: Bill 1857; Bill 1903:46; Bill 1906:62; Bill 1948:34; Bill 1978/79:110, 12; Bill 1992/93:226; Bill 2007/08:108; Bill 2013/14:141; Swedish Forest Agency 2005. Table adapted from Beland, Lindahl *et al.* (2015). Graphical design: Jerker Lokrantz/Azote.

resulting in an 'implementation deficit' (Beland Lindahl *et al*. 2015). Hence, a considerable disconnect arises between the plethora of ambitious, formally expressed policy goals, and real-world forest management practices. Instead of integration, the goal-oriented forest policy results in a growing gap between what is desired and what is delivered.

Moreover, the establishment of goals, and efforts to reach them, has problematic temporal and spatial dimensions. The strong future orientation of the goals raises major problems about how futures should be represented and by whom, and the specific solutions in the present that are required for this (ideal) 'world making' (Brown *et al*. 2000; Andersson and Keizer 2014). Thus, the goals may impose a narrow and closed notion of future developments, rather than widening the present-day focus and providing opportunities for alternatives and creative resolutions (Stirling 2008; 2010). Likewise, the translation of global goals to national goals and practical applications has often resulted in specific local-level targets and implementations. To implement these in a deregulated context, voluntary undertakings, participatory processes, and certification have evolved. However, since the goals are already set, such processes become ways to persuade actors to implement predetermined solutions, rather than democratically to create collective pathways to handle the challenges. Consequently, there has been fierce debate about associated threats to property rights, lack of respect for traditional values created through efforts of generations of forest owners, competing land-use interests, and the inability of state authorities to leave things be (Zaremba 2012; Sténs *et al*. 2016).

Reflexive Forestry

How, then, should we instead proceed? As we will see in Chapter 7, several integrated approaches intended to bridge natural resource government and management have been proposed and applied. However, many of these integrated approaches tend to prioritize objectives (e.g. ecological integrity or resilience) that may not be fully compatible with production-oriented systems, such as Swedish forest models (Rist *et al*. 2013). We therefore argue that there is a need for a new approach that may be applied in a production-oriented system with the aim to promote sustainable development.

We call this new integrated approach 'Reflexive Forestry'. At first glance these terms could seem contradictory. 'Reflexivity' is usually used in social sciences and humanities when referring to critical and social process-oriented understandings of ambiguous transitions (Beck 1992; Beck *et al*. 2003; Voß *et al*. 2006). Reflexivity means both 'self-confrontation' (how societal processes give rise to unintended risks and uncertainties that confront society and must be acted upon), and 'reflection' (the constant construction of knowledge by individuals and organizations) (Boström *et al*. 2016). In contrast, 'forestry' is associated with rationalistic and instrumental processes, rooted in natural sciences, and management oriented towards meeting desired goals, needs, and values for human and environmental benefits (Dargavel and Johann 2013). While reflexivity is a starting point for initiating critical reflection, dialogue and deliberation, forestry is an end in itself, applying knowledge-based measures to achieve certain real-world objectives.

Thus, Reflexive Forestry combines a reflective and analytical tradition of 'questioning and diagnosing' with a positivistic and action-oriented tradition of 'knowing and doing'. These traditions are not wholly incompatible. Indeed, integrating them in a coherent process is essential to enable the emergence of forest governance and management capable of delivering sustainable development during a transition with numerous uncertainties.

Moreover, Reflexive Forestry is a matter of handling the dilemma of 'opening up' and 'closing down' social problem-oriented processes (Voß and Kemp 2005; Leach *et al.* 2010). In order to produce robust knowledge and develop a wide spectrum of potentially feasible policies and choices to consider, problem-oriented interactions need to be open, inclusive, and transparent to take account of diverse understandings, factors, and interests. However, to enable timely decision-making and action, options and inputs must also be closed down at appropriate points, through processes that reduce complexity, identify the most relevant factors, and establish stable resolutions, including risk assessment, cost-benefit analysis, and stakeholder negotiation. Thus, efforts to open up processes must be complemented with strategies to close them down. Hence, Reflexive Forestry is a process to direct forestry activities while reflecting, to involve conflicting interests and actors while finding collective pathways, to perform while learning, and to pursue aims while adjusting courses.

Aim and central themes of the book

The overarching aim of this book is to explore historic and future outlooks, as well as contemporary integrated approaches, in both environmental governance and forest management. This exploration is centred on empirical foundations drawn from the Swedish forest arena. Based on the findings we develop an integrated approach, Reflexive Forestry, to facilitate attempts to close the identified gap between governance and management. As described in subsequent sections, this involves ways of thinking and progressing into action that coherently connect knowledge, benefits and understandings, diverse social levels and actors, and both practices and outcomes. Key conceptual elements of this approach, and thus central themes of both this book and Reflexive Forestry, are as follows.

Transtemporal thinking emphasizes that, while it is essential to extend our outlook backward and forward in time, it is also vital to connect broad temporal understanding to specific social processes, decision-making and peoples in our time. Rhetorical remarks about 'learning from the past' and our 'responsibility for future generations' are legion. As these remarks indicate, time is generally seen as linear and progressive, with clear separations between the past, present, and future. Koselleck (2004) uses the term 'temporality' to capture how the experiences of time have changed historically, including notions of progress, decline, acceleration, and timing. How we apprehend time, and the relations between past, present and future, affects our understanding of possibilities to influence ongoing transitions (Hartog 2015). For example, in the global warming debate, graphs covering huge time-spans have had tremendous impact (Mann 2013). Nevertheless, without questioning the scientific credibility of the graphs (and underlying studies), apprehension of such mega-trends tends to reduce the present to a momentary intermediate stage between the past and the future. Thus, such 'big pictures' tend to diminish our apparent power and possibilities to influence future developments.

Transtemporal thinking, as defined in this book, acknowledges that the past and future are inextricably linked to the present, constantly influencing our decision-making in real time (cf. Adam 2013; Fareld 2016). This understanding differs from a traditional cross-generational mode of thinking that has dominated scientifically based forestry for the last three centuries and the sustainable development agenda for the last three decades (Warde 2011; Nordblad 2016). In contrast to these progressive, linear and future-oriented notions, transtemporal thinking is contextual, nonlinear and characterised by presentism. Hence, we recognize well-informed and responsible decision-making in the present as the final step in a broader appraisal of forests and forestry in their temporal context.

Ways of knowing are methods of acquiring knowledge about the universe, our world, and their myriads of interacting constituents, by reducing the complexities involved sufficiently to make sense of the general patterns and principles. Pickstone (2000) divides scientific ways of knowing into: world reading (interpretation), sorting (describing and classifying), analysis (breaking objects of study into pieces), synthesis (bringing elements together), experimentalism (testing), and technoscience (know-how and applications). Forestry sciences entail all these ways of knowing, but the technoscience-based elements generally focus on economic and political aspects; the underlying values and understandings that tie science, policy-making and commerce together, and direct the ways of knowing and innovations along certain tracks (Pickstone 2000). Consequently, forests and forestry can potentially provide bridges between forest-related science, economics and politics. This ability is pivotal when trying to integrate forest governance and management, and to achieve change. However, instead of building tight partnerships with a restricted number of actors, which has often been done in the past, we emphasize the importance of collaborations fostered by building networks with a multitude of actors.

Similarly, forest sciences have both theoretical and practical elements, so it is important to incorporate other forest actors' ways of knowing, such as local forest owners' and users' practical knowledge. Human societies all over the world have developed experiences and explanations relating to their environments. These ways of knowing are often referred to as traditional ecological knowledge, or indigenous or local knowledge (Berkes 1999; Folke 2004; Rist *et al.* 2010). This knowledge has been the basis for diverse activities that sustain societies in many parts of the world, and was recently acknowledged in global governance initiatives, such as the UN's Convention on Biological Diversity (CBD 1992) and 2030 Agenda for Sustainable Development (European Commission 2015), as an important factor to enhance sustainability. We therefore endorse the establishment of networks with diverse actors and the development of interactive and participatory methods to integrate different ways of knowing when addressing urgent issues.

Ways of doing (in this context) are modes of forest governance, methods of forest management, and practices. For instance, forest governance can be implemented by top-down or bottom-up processes, state- or market-driven, and regulated by soft or hard law. Similarly, forest management can be conducted by one-size-fits-all or adapted solutions, by experts or participatory processes, and by focusing on technological fixes or seeing management in a context of wider considerations. In democratic societies with private property rights, a public sphere, and market economy, forest governance and management require navigation in a context where power is disseminated across diverse societal subsystems and among many actors with different ways of knowing, notions, and desired forest benefits (Hooghe and Marks 2001; Liesbet and Gary 2003; Eckerberg and Joas 2004).

To handle such polycentric conditions, it has become increasingly clear that ways of doing, governance, management and practice, need to be conducted in interaction with multiple actors (for example, local actors and citizens) through participatory and pathway-developing processes mediated by various kinds of dialogue (Leach *et al.* 2010). Thus, seen from this perspective, the ways of doing must not be restricted to implementing, disseminating, and applying certain forms of governance and management in society. In addition, they must enable mutual interactions and learning across social levels. Overarching dilemmas and opportunities must be identified and addressed in different socio-ecological contexts, and all the diverging responses should be considered in attempts to resolve the dilemmas and grasp opportunities. Hence, forestry measures must be adjusted to relevant situations to become applicable, and involving diverse actors with equally diverse motivations may enable engagement and creativity that yield powerful innovations, examples of best practices (or illuminating failures), and other potent feedback.

Overall, Reflexive Forestry is about building up the whole society's capacity to take care of and develop a vital natural resource. Accordingly, we see forests as constituting a *social arena* composed of: the associated forest resources and benefits; the actors, organizations, institutions, and markets involved; and their understandings, norms, ways of knowing, ways of doing, and motivations. We use the term 'arena' instead of the usual term 'sector', because an 'arena' plays vital social roles, and is open to the whole society and diverse actors, in contrast to a 'sector', which divides society into separate parts with selected actors. Arenas are also sites of competition and challenges associated with conflicting claims, coalitions, and interactions among the actors. Importantly, to create an interplay between the actors in an arena there must, to some extent, be common rules for coexistence and development. These include both explicit regulations and implicit norms, which constantly formulate and reformulate what is seen as acceptable and unacceptable, and desirable and undesirable behaviours and ambitions.

Hence, through competing claims and understandings and through mutual cooperation, compromise, and commitments by each party involved, a common order and directionality may be generated and sustained. Such broad societal agreements, or what we call (in this context) *forest social contracts*, are sets of implicit and/or explicit arrangements built on trust and relational commitment between involved actors to achieve certain objectives. The establishment of a forest social contract does not mean that underlying conflicts disappear. When the situation changes, and new understandings and claims arise, the legitimacy of the contract erodes, opening opportunities (or necessities) for renegotiation or possibly the development of a new contract. Using the term 'forest social contract', thereby connecting to the contract paradigm of classical political philosophers such as Thomas Hobbes, John Locke and Jean-Jacques Rousseau (Boucher and Kelly 1994; Skryms 2014; Messner 2015; Muldoon 2016), enables us to raise questions about legitimacy, rights and duties, justice, and use of power. It also implies that relations between the rulers and the ruled, and between individual and collective action, should be considered.

The Swedish forest arena

In this book, we use information drawn from the Swedish forest arena as an empirical base to address forest governance and management challenges. The aim is not to comprehensively cover or summarize all aspects of the Swedish forest arena. Instead, we focus on conditions, themes, and issues that are relevant for our general arguments regarding the lack of integration between forest governance and management, and the development of Reflexive Forestry.

In relation to its size, population, and share of global forest assets, Sweden currently has disproportionately large importance in international forestry (FAO 2010; Dauvergne and Lister 2011). In Sweden, property rights are strong, in international terms (The International Property Right Index 2016), and the private forest owners are key actors in the forest arena (Stjernquist 1973). Nevertheless, most of the Swedish forests are open to everyone and they provide multiple benefits, due to the Swedish tradition of right of public access to private land (*allemansrätten*) and other regulated usufructuary rights, so other interests must be considered (Sténs and

Sandström 2013). Hence, forests and forestry are of national importance for everyone, rather than merely those who have property rights or who are directly dependent on forest resources.

The Swedish forests cover about 28 million hectares (about 70 per cent of the country's total land area), from the temperate zone in the south to the boreal zone in the north (SFA 2014). Approximately 75 per cent of the forestland is under active management for multiple purposes, while the rest is subject to management restrictions, particularly concerning timber production. About 250,000 small-scale forest owners hold about half of the forestland. The rest is owned by large forest companies, the government (through the state-owned forestry company Sveaskog and the National Property Board, Fastighetsverket), and various townships and parishes. Moreover, forestry accounts for 2.2 per cent of Swedish GDP and the forest industry accounts for 9–12 per cent of the employment, exports, turnover and added value in Swedish industry. The forest industry is strongly export-oriented and more than 80 per cent of the forest-based products are exported, placing Sweden among the world's largest exporters of these products. Although forestry clearly plays a substantial role in Sweden, its relative importance has decreased. For example, in the 1920s and 1930s about 50 per cent of the country's export revenues came from forestry (Siiskonen 2013). In addition, 150 years ago 90 per cent of the Swedish population lived close to forests in the countryside, while 85 per cent now live in cities (SCB 2015). Thus, the forest social arena is not constant, but dynamic; it has undergone several transformations in the past, and it will continue to change in the future.

Due to the forest industry's long-term importance, there are well-developed regulations and institutions to support the arena. Thus, it is a public sphere inhabited by politicians, forest owners, researchers, professionals, businessmen, journalists, activists, and citizens. Numerous organizations also engage in the arena (to varying degrees). These include: state authorities; forest companies; non-governmental organizations; national trade and employers' organizations linked to the pulp, paper, and wood-processing industries; regional cooperative associations of small-scale private forest owners (which collectively have hundreds of thousands of members and commonly owned sawmills and paper and pulp factories); associations of forestry contractors; trade unions; forest research institutes; and other professional associations.

The environmental and nature protection movement has spawned another important group of organizations that participate in interactions in the arena, seeking to address concerns about nature conservation, environmental issues, and biodiversity. These include international organizations such as the World Wildlife Fund (WWF) and Greenpeace, as well as national organizations, such as the Swedish Society for Nature Conservation, local groups and transboundary activist networks. Several other associated sectors and groups (including energy, recreation, tourism, hunting, agriculture and the

Saami reindeer herding communities) are also represented by influential organizations in the arena. All the mentioned organizations also have global connections and are associated with (or members of) international organizations and corporations (Dargavel 2010). Finally, Sweden has a tradition of a strong state governing with independent national authorities with regional and local offshoots. Thus, forests and forestry constitute a societal arena that engages numerous members, representatives or agents of (and plays vital roles in) the state, trade and industry, and civic society.

The Swedish forest arena is used to exemplify and illuminate the more general issues and arguments in this book with wider implications and relevance to other forest and decision-making contexts globally. To some extent, the Swedish forest arena is similar to forest arenas in other countries with active forest management, and faces similar challenges regarding forest owner structure, multiple goals and the increasing numbers of trade-offs needed. Moreover, the Swedish case reflects multi-level associations that affect decision-making from the establishment of global conventions, through transnational governmental forms (e.g. the European Union) and national policy-making, to forest owners' local-scale management decisions. We also aim to situate the Swedish forest arena in a context of relevant international research on forest history, future studies, silviculture, nature resource management, governance, and integrated approaches. Thus, a wider ambition is to connect our observations and conclusions to current issues, and to contribute to ongoing discussions, in these fields.

Outline of the book

This book is divided into four parts (Figure 2). The first three concern the dimensions of transtemporal thinking. These temporally oriented parts are not structured in a classical chronological (past, present, and future) order. Since we argue that both historical and future dimensions influence present situations, states, conditions, and actions (while perceptions of history and the future are always influenced and understood from current perspectives), the order is instead: past, future, and present. Finally, after presenting these dimensions (and their implications), Reflexive Forestry is elaborated.

The first part, 'Looking back', explores historical layers of forest benefits, knowledge, and governance that have accumulated in the forest arena's foundations during the last 150 years. This includes the development of forest benefits, associated uses and rights, the rise of scientific forest knowledge and management approaches, and the development of forest regulation and governance. This part shows how forest social contracts have evolved, successively replaced previous contracts, and collectively created overlapping objectives and structures. Thus, it problematizes how complex and dynamic historical dimensions feed into present-day trade-offs and forest policy decision-making.

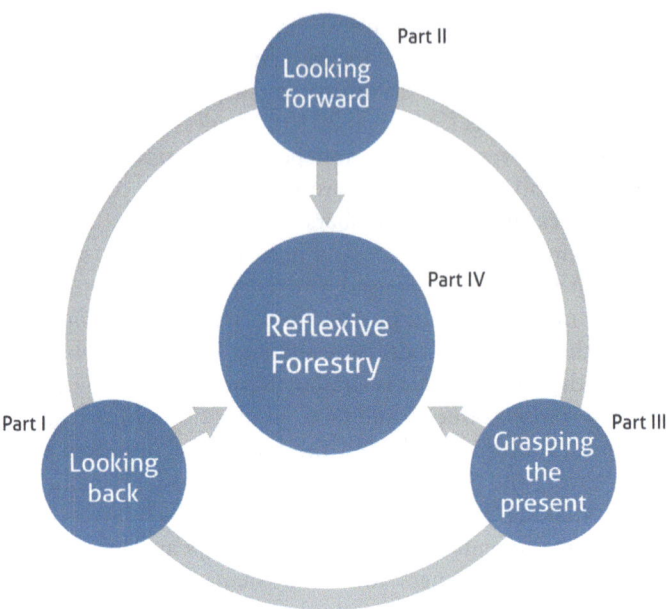

Figure 0.2 The structure of the book.

Graphical design: Jerker Lokrantz/Azote.

The next part, 'Looking forward', considers how alternative forest futures can be assessed, and how they may inform policy-making and management in the present. While we all have experience of history, and can empirically investigate past events, we have no experience of the future and it cannot be empirically researched by any currently conceivable methodology. Nevertheless, attempts to foresee the future are essential for a society to anticipate possibilities and challenges, as well as to plan for sustainable development. This section assesses an array of techniques – quantitative and qualitative methods such as system analysis, scenario analysis, backcasting, and trend (trajectory) analysis – that can be used to explore possible, probable, or preferable futures. This part also discusses what it means to make future-informed choices and trade-offs in the present, by taking responsibility for the future consequences of aspirations and projects, and thus how better notions of alternative futures foster greater reflexivity. Thereby, a better notion of alternative futures enables reflexivity.

The third part, 'Grasping the present', compares several integrated approaches, all intended to enable change, learning, and common action to promote sustainability in a transforming and multi-layered present. In modern society power is distributed across numerous actors, systems, and levels, which complicates such efforts, and is related to the increasing demands on services provided by forests, such as social values – and climate change mitigation. This part examines and compares integrated forest governance and management approaches to deal with trade-offs among ecosystem services and competing claims from different actors, and implementation of management decisions. It also analyzes how risk and uncertainty may be considered in policy making.

Finally, in the last part, the Reflexive Forestry approach is elaborated by outlining principles and tools that can integrate forest governance and management, and guide efforts to develop reflective, deliberative, and tailored forest management.

Part I
Looking back

Revealing historical layers

The Swedish forest arena has evolved in a European context where a large proportion of the forests are privately owned, limiting the state's capacity to steer private owners' use of their land. However, unlike most of continental Europe, Swedish forest governance and management has developed in a setting where forest-related industries have been essential for national economic growth. Thus, strong forest industrial production interests have dominated the forest arena and economic objectives have been prioritized. Profitability and evenness of revenues have been guiding principles (Brukas and Weber 2009) and managers normally apply standard investment analysis techniques in scheduling silvicultural activities, typically using a real discount interest rate of 2–3 per cent (Simonsen *et al.* 2010). Land ownership, prioritized benefits, dominating power-relations, established traditions, and developed governance models have co-evolved, collectively shaping a high degree of path-dependency. Consequently, the Swedish forest arena is constrained by its institutional past, which hampers present efforts to promote required change (Beland Lindahl *et al.* 2015).

However, there have been clearly discernible shifts during the Swedish forest arena's history, reflecting changes in circumstances, understandings, and ambitions. During the last 150 years, new actors have entered the arena, additional forest uses and rights have been promoted, new knowledge horizons have opened, and novel forest governance and management models have been introduced. A series of replacements of broad forest social contracts with new contracts, introducing new goals, benefits, frames, and stakeholders after periods of enhanced debate and changing conditions, can also be discerned. This indicates that transformative change is driven by a combination of conflicts and crises, which create external pressures and internal incentives for the reorientation of the arena in order to maintain its competitiveness, political influence and/or legitimacy.

Thus, the forest arena's historical development has been dynamic and complex, leaving a similarly complex legacy of various historical layers feeding into the present time, both imposing inherent constraints and raising possibilities for change. This section explores in thematic chapters three historical layers that must be grasped to understand the arena's development. These chapters chronologically cover the last 150 years from the start of the modern forest industry until the present time. However, the aim is not to comprehensively describe the whole Swedish forest history, partly because we are more interested in illuminating patterns and partly because there are already many broad studies covering various aspects of forest history during this period (e.g. Stjernquist 1973; Eckerberg 1990; Linder and Östlund 1992; Östlund 1993; Hytönen 1995; Östlund 1997; Östlund *et al.* 1997; Eliasson 2002; Lisberg Jensen 2002; Jörnmark 2004; Kardell 2004; Lehtinen *et al.* 2004; Appelstrand 2007; Enander 2007; Josefsson 2009; Antonson and Jansson 2011; Simonsson 2016).

More specifically, the first chapter of this part deals with the development of the forest arena and trade-offs between forest benefits and their relations with ownership and various forest rights. The second chapter focuses on the emergence of scientific forest knowledge and management approaches, and how new knowledge has changed understanding of forests' roles. The third chapter deals with the evolution of forest governance, and how broader social agreements – forest social contracts – have been successively replaced with new contracts. The overall goal of this section is to develop a reflexive understanding of the present situation by elucidating these historical layers.

1 Forest benefits

1.1 Introduction

In Sweden, it is generally taken for granted that forestry is a vital industry providing benefits to society as a whole. Forestry was as an important part of the industrialization processes in the nineteenth century, essential for establishment of the welfare state in the mid-twentieth century, and a vital resource for multinational companies and other players in the global arena during the last half century (Antonson *et al.* 2011). Thus, forest benefits were tantamount to profit and economic growth.

However, the importance of 'the green gold' represented by forest resources, and the forest industry, is not self-evident. Instead, the emergence of this importance needs to be understood as a historical process, in which the industrial value of the forest resources developed in relation to other benefits. While the forest industry has mainly used the forest resources to produce timber, pulp and paper, there are other benefits from the forest, such as energy sources, food, handicraft, recreation, tourism and nature conservation, that have been developed in parallel, with varying tensions. With diverse owners, ranging from small-scale private owners (often also involved in agriculture) to large-scale forest companies and the state, there have also been tensions between different interests and understandings. Nevertheless, forests do not only benefit their owners. On a more individual and local scale there are long traditions of diverse usufructuary rights, covered by various regulations, some of which are included in the Swedish Right of Public Access (*allemansrätten*) and others in indigenous rights. Moreover, a constant issue has been the tension between private property rights and the forest's treatment as a common good benefiting the whole society both through its economic importance and wider environmental significance.

In this chapter, we trace the development of forestry as a strong industry in relation to other benefits and rights in the forest arena and shifting historical contexts. The chapter starts by outlining the development of Swedish forests' ownership structure, and the transition of forestry from a part of subsistence farming and proto-industry to a major commercial industrial sector. It continues by describing the evolvement of other forest-dependent benefits and rights, then ends by focusing on the 'environmental turn' in society and how it has affected forestry during the last decades.

1.2 The development of forest ownership

Property rights have played a fundamental role in formation of the forest arena. The forest social contracts include formal and informal regulations regarding rights and duties of the forest owners, state and other actors. The composition of ownership has been relatively stable since the early twentieth century, but is the result of several land reforms with different aims in different times and parts of the country (Antonson and Jansson 2011). Forest ownership in the more densely populated south, Götaland and part of Svealand (Map 1), has been regulated since the Middle Ages. Private landowners have long dominated in this part of the country. The forested northernmost areas (Norra and Södra Norrland, and part of Svealand) became strongly regulated by the central authorities much later, and are now dominated by the private sector and state-owned companies.

Boundary delimitation processes that occurred in the northern parts of Sweden between 1683 and 1920s played a major role in creation of the current ownership distribution. The crown and ironwork companies, which had dominated Swedish export industries since the Middle Ages, initiated these processes, in which private land (including all forest that was not formally owned by someone) was claimed by the state to secure access to charcoal. Large proportions of state-owned forests were then leased and subsequently sold to the ironwork companies. Between 1780 and 1860 large areas of forestland were transferred to private companies in this manner. In the northern inlands, the delimitation was resisted by settled farmers and the Sami, who used forests as extensive grazing lands. Consequently, new regulations, which favoured the farmers more, were passed at the end of the eighteenth century and the delimitation processes continued (Jörnmark 2004; Lundmark 2008; Lundmark and Rumar 2008).

Generally, during the nineteenth century private property rights became more definite in Sweden, as in other parts of Europe. This boosted the delimitation processes and gave landowners greater leeway to use, sell, mortgage, parcel-off or lease out land (Morell 2011). The development was much less beneficial for the increasing numbers of landless people who became more dependent on private landowners. The delimitation processes also marginalized the Sami who had usufructuary rights regarding reindeer husbandry and other forest uses in northern parts of the country. The Sami are an Indigenous people who have practiced reindeer herding since 'time immemorial' across Sápmi – the traditional Sami homelands that currently comprise parts of Sweden, Norway, Finland and the Kola Peninsula in Russia. Sápmi covers approximately 50 per cent of Sweden and includes 40 per cent of productive forestland (Sandström 2015). The authorities' disregard of the Sami during the delimitation processes can be partly

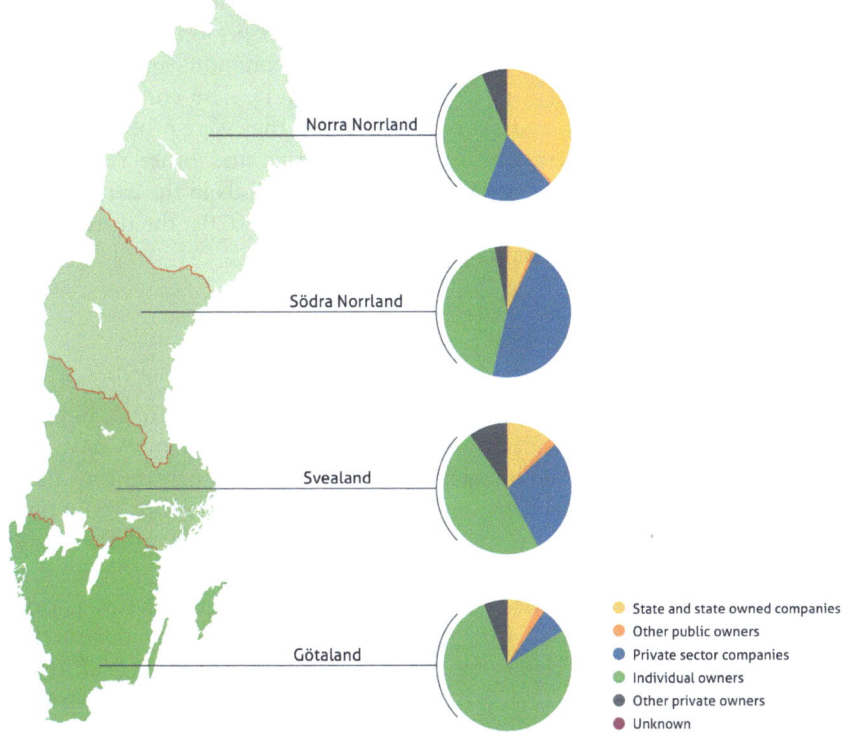

Figure 1.1 Shares of productive forest (hectares) by indicated categories of owners, in the Swedish regions in 2012.

Source: *Swedish Statistical Yearbook of Forestry 2014*, table 2.7. Graphical design: Jerker Lokrantz/Azote.

explained by the expansion of new industries in Sápmi at the time. Thus, the boreal forest was gaining new commercial value of interest to the state, which competed with Sami culture and businesses (Lundmark 2008).

In the mid-nineteenth century, the value of timber steadily increased, mainly because of demand from the British market. At this time, many of the Swedish ironwork companies, which already owned large areas of forest, moved into or merged with timber and pulp businesses. Indeed, all of today's multi-national Swedish forest companies originated as ironwork industries, some in the Middle Ages (Jörnmark 2004). As it became easier to trade land and the value of wood products increased, companies started to buy harvest rights, forests, and entire estates from private landowners. This process was so extensive that it came to be regarded as a major societal problem in the late nineteenth century, destabilizing rural livelihoods and forest yields. Thus, to halt these changes the state allocated forestland to groups of farmers and settlers for use as 'forest commons' from 1861, mainly in northern parts of the country (Kungl. Skogs- och lantbruksakademien 2011).

However, the companies also managed to acquire shares in commons, and private forest companies now possess over 50 per cent of the shares in some of these privately owned commons. The commons are included in the category of 'other' private owners (Figure 1.1). The companies kept expanding their forest ownership until 1906, when they were prohibited from buying forestland from private owners. With this and other regulations launched to safeguard both forests and rural livelihoods in the early 1900s, and completion of the delimitation processes in the 1920s, the distribution of forest ownership was stabilized.

1.3 From agricultural 'appendage' and proto-industry to industrial forestry

Strong property rights are fundamental aspects of the forest arena, but they are not unrestricted. Instead, as already mentioned, non-owners have had various rights (which have changed over time) to access private forestland and enjoy certain benefits. From about mid-nineteenth century to about mid-twentieth century, there was pervasive tension in benefits of the forest between agriculture and industry. In the rural communities, where more than 80 per cent of the population still lived in the beginning of this period, forestland was mainly used for grazing and as a source of construction material and firewood, except in regions with ironworks (Östlund *et al.* 1997; Kardell 2004; Kardell 2016). Use of the forestland was not strictly restricted to the owner, as tenants and crofters had usufructuary rights to collect essential goods for consumption or small-scale trade, such as fuel, berries, herbs, game and handicraft materials. Hence, for most of the population the forests were mainly valued for their importance to agricultural and daily livelihoods. Consequently, a recurrent complaint among foresters in the nineteenth century was that forestry was regarded as an 'appendage' to agriculture and not as an independent industry in its own right (Falkman 1852; *Underdånigt betänkande* 1856).

The forest arena changed with industrialization and new ways of doing. By the late eighteenth century, proto-industrialization by farmers expanded with production of timber, tar, charcoal, and potash. In some areas, such activities were practiced on a large scale, resulting in considerable export revenues for Sweden, peaking in the nineteenth century (Kardell 2004). Large-scale industrialization started with establishment of the first steam-powered sawmills in the mid-nineteenth century. This followed a wave of establishment (from south to north) of sawmills along the coast of Gulf of Bothnia, which had tens of thousands of employees, and subsequently pulp and paper mills. In the decades around the turn of the twentieth century, the Nordic countries were the largest exporters of wood products in the world (Williams 2003). This development led to the formation of a national industrial cluster involving industrial actors, the state and other stakeholders (Sölvell *et al.* 1992), a powerful network of actors that strongly influenced

Figure 1.2 Cows pasturing on site that had been sown twelve years earlier, consequently damaging the regeneration.

Photo: Lars Tirén 1940, Nyby, Lycksele, in Västerbotten. Copyright Swedish University of Agricultural Sciences.

national forest policy, regional development and initiation of the forest arena as a whole.

This development continued in the twentieth century, and took a new turn in the middle of it. In Sweden, the forest industry and exports of timber and paper boomed as international demand for construction material soared after World War II and during the Korean War in the 1950s (Bernes and Lundgren 2009). This led to strains to meet the demand and a shortage of workforce in the labour-intensive forestry, met by rationalization and mechanization of forestry to increase production. The establishment of high-productivity forests was prioritized through intensive management, involving clear-cutting of large tracts, wetland drainage, and replanting with monocultures, sometimes of exotic tree species (Mårald and Westholm 2016).

The forests were thus designed to serve as sources of raw materials to support the forestry industry. An important requirement for mechanized forestry was cheap imported oil from the Middle East. Overall, the changes eased the old competition between agricultural and industrial use of the forest assets, and the separation of land uses (Morell 2011). Moreover, the energy-shift from wood to imported oil sharply reduced needs for firewood and fuel, enabling the clear prioritization of timber production (Egan Sjölander *et al.* 2014).

1.4 The regulation of multiple forest activities and benefits

Despite the shift towards more intensive industrial use of the forest resource, other forest ways of doing and benefits have continued and co-evolved in parallel. They include both traditional activities connected to agriculture and reindeer herding, and new activities such as nature conservation, tourism, and recreation. This has resulted in complex and overlapping regulations concerning various aspects and benefits of the forest landscape. Moreover, it has shaped complex relations between different rights, including (among others) private property rights, the Right of Public Access, the Reindeer Management Right and wider public interests.

In the early twentieth century voices were raised in favour of preserving scenic and untouched areas of nature. In 1909, the year the Swedish Society for Nature Conservation was founded, the Swedish parliament passed the Nature Conservation Act to preserve unique natural monuments and national parks (Lundgren 2009). Nine national parks were established at the time, most in the mountains of Lapland, a region regarded at the time as having no economic value, so parks created there would not restrict forestry or mining (Ödmann, Bucht, and Nordström 1982). As in other nations around the world, national parks have been treated as national treasures, symbolizing a pristine and vanishing nature. They have also been set aside for scientific reasons, as well as for tourism and recreation (Lundgren 2009).

At the time of writing, twenty-nine national parks have been established (SEPA 2016a). Moreover, the parks established by the Conservation Act of 1909 were subsequently augmented by the creation of new legal institutions, such as: 'nature reserves' (in 1964); 'wildlife refuges' (in 1965); 'nature conservation areas' (in 1974); and 'protected habitat areas', 'Natura 2000' protected areas and 'voluntary set asides' (in the mid-1990s). Today, conservation areas cover about 13 per cent of the total land area in Sweden, still strongly concentrated in the mountain areas. Around 7 per cent of the forestland is formally protected (i.e. excluding voluntary set asides), including 4 per cent of productive forests (SEPA 2016b). Conservation policy has been heavily based on the idea of protecting nature from local people and their everyday requirements. The government has strived to delineate the protected areas from the surrounding areas, by top-down management and a low degree of involvement of local actors (Zachrisson 2009; Hovik *et al.* 2009).

Following the Great Depression of the 1920s and 1930s, and world wars, welfare states expanded in the Western world and, among many other changes, there emerged new forms of forest uses, understandings and benefits. In Sweden this development took off in the 1930s. With eight-hour working days and two weeks of statutory holiday for all employees, there was increased debate about the need to establish useful and wholesome forms of recreation for all citizens. As a response, easily accessible open-air recreation areas, often including forests, lakes, and mountains, were established (Sandell and Sörlin 2000). In this context recreation and tourism were supported by the Swedish traditional right of public access to privately owned forests and other nature areas. However, the historical background and meaning of the concept are contested, and the right has never been defined in detail by law. Instead, it is restricted indirectly by other laws and regulations, e.g. the Real Property Law (SFS 1970:994), the Criminal Code (SFS 1962:700) and the Environmental Code (SFS 1998:808). However, in 1994 the Right of Public Access was enshrined in the Swedish constitution (Sténs and Sandström 2013; 2014), in very generous terms. It stipulates that the public may roam on almost all land, irrespective of ownership, as long they do not interfere with the landowner or cause any damage to the land and what is on it. It also provides leeway for the use of berries, mushrooms, common wild flowers, and herbs (Åslund 2008).

Another forest benefit, not included in the Right of Public Access, is hunting, especially moose-hunting, which is a major social movement in Sweden with nearly 300,000 hunters and strong organizations (Sandström *et al.* 2013). In Sweden, the moose management regime has changed several times since it was first introduced in the early twentieth century (Wennberg DiGasper 2008). The moose population exploded in the 1960s, due to a combination of regulatory changes (new hunting legislation), and the rationalization of agriculture (livestock were no longer kept in the forests for grazing) and forestry (the introduction of clear-cutting increased food for moose) (Sandström *et al.* 2013). When the moose population peaked in the 1980s, almost 200,000 moose were harvested annually (Lavsund *et al.* 2003). Moose cause severe browsing damage to forest plantations, so moose management has been a source of conflict between industrial forestry and hunting interests. Therefore, stakeholder participation and cooperation in Swedish moose management has been reinforced, over time, by devolution of management rights to more local levels and the management system has become more adaptive to such influences (Wennberg DiGasper 2008; Bjärstig *et al.* 2016).

Reindeer husbandry is also practiced in parallel with forestry and (to much lesser extents) other activities (Sandström and Widmark 2007; Widmark 2009). The right to keep and herd reindeer is settled, exclusively for the Indigenous Sami, in the Reindeer Husbandry Act (1971:437). The right is also protected in the Swedish constitution and considered a strong usufruct based on rights prescribed since time immemorial (Allard 2016). In order to reduce conflicts between the two sectors, consultation procedures were introduced by the Swedish parliament in 1979, and about twenty years later they were extended to cover a larger geographical area through the certification system run by the Forest Stewardship Council, FSC (2009) (Swedish Forestry Act, 1979:429; Swedish Reindeer Husbandry Act, 1971:437).

However, the consultation procedures do not seem to fulfil their purpose, since conflicts between the two actors are still ongoing. While forest companies and small-scale forest owners are the owners of the land, reindeer herders have legal usufructuary rights to use it for reindeer herding, hunting and fishing. However, reindeer herders have difficulty claiming these rights. The laws and legal procedures regulating the relationships between the two sectors do not seem to give sufficient protection to the natural grazing areas needed for reindeer husbandry, creating an imbalance in property rights (Swedish Reindeer Husbandry Act, 1971:437; Hahn 2000; Widmark 2009; Sandström 2015). Efforts have been made to identify solutions to this conflict by developing tools, techniques and collaborative procedures (Johansson 2013; Bostedt *et al.* 2015; Horstkotte *et al.* 2014), but as long as there are unresolved issues of property rights there will still be room for conflict between industrial forestry and reindeer husbandry.

1.5 From the environmental turn to bioeconomy

Modern understanding of the environment shifted substantially between about 1950 and the early 1970s (Warde and Sörlin 2015), as emerging global statistics showed that development of modern society was being accompanied by similar rates of population growth, soil erosion, deforestation, water shortages, depletion of resources, industrial pollution and species losses. The resulting view of the development as a degenerative process pointed towards a catastrophic future – a transtemporal idea. These diagnoses of the environment, the direction of change, and the prediction of the future entailed a new reflexivity on the human condition (Warde and Sörlin 2015; Mårald and Westholm 2016).

From the 1960s onwards the emergence of environmentalism had far-reaching effects on forestry and the forest arena, raising doubts about the sustainability of prioritizing high production and financial profits. In Sweden, a central element in this critique was 'the chain of large-scale forestry'. That is, the idea that heavy mechanization, clear-cutting, soil scarification, even-aged monoculture plantations, and use of introduced tree species, fertilizers, and chemicals were all parts of a closely connected system (Fältbiologerna 1973). Thus, according to this critique, the prevailing clear-cutting forestry regime entailed a whole package of ecologically harmful methods. In the aftermath of Rachel Carson's *Silent Spring* (1962), the use of herbicides was particularly condemned, and after a long and heated public debate their use was prohibited at the end of the 1970s (Lisberg Jensen 2006; Simonsson *et al.* 2013).

In addition, other intensive forest management methods were gradually withdrawn or modified. In the early 1980s the use of fertilizers sharply decreased, partly in response to rising public concerns (Lindkvist *et al.* 2011) that were connected (*inter alia*) to alarm in Germany about reductions in forest growth and risks of major forest dieback (*Waldsterben*) due to atmospheric pollution (Radkau 2012). Soon signs of similar dieback in Southern Sweden led to many newspaper headlines and a major scientific

Figure 1.3 Spraying with herbicides to control aspen in a pine plantation. 1953, Kulbäcksliden, Västerbotten.

Photo: Lars Tirén. Copyright Swedish University of Agricultural Sciences.

controversy (Tunlid 2007; Lidskog and Sundqvist 2011). There were also controversies about forestry in mountainous locations and the threats posed to the last remains of old growth forests (Lisberg Jensen 2002). Moreover, in northern Sweden extensive plantations of North American lodgepole pine (*Pinus contorta*) were established in the 1970s and 1980s to increase volume production to be able to avoid future shortages of wood. However, in the mid-1980s reports of widespread fungal damage to lodgepole plantations led to much publicity, and subsequently to harsher restrictions in the locations and sizes of lodgepole pine plantations (Elfving *et al.* 2001; Backman and Mårald 2016).

Overall, in the 1980s and early 1990s forest issues were acrimoniously debated. This reflects a global trend with similar conflicts in many other regions. For instance, 'the timber war' in the USA challenged the US Forest Service's focus on timber production and the management of federal lands (Winkel 2014). Consequently, timber production on national forestland sharply declined, by 85 per cent, between a peak harvest in 1987 and 2000 (Mårald *et al.* 2016). With less focus on timber production, forest management turned to ecological restoration and meeting new objectives and challenges, including increased demands for clean air and water, wildlife habitats, and opportunities for outdoor recreation.

Similar trends occurred in Sweden. The Swedish nature conservation policy broadened to include economic objectives related to nature-based tourism as a basis for local development, as well as social (tourism and outdoor recreation), cultural (local identity), ethical and moral (nature's intrinsic value), and scientific objectives (Overvåg *et al.* 2016). Changes in nature conservation policy have also facilitated (with mixed results) the establishment of partnerships between the government and private forest owners to protect forestland in order to preserve both biodiversity (Widman 2015) and social values (Widman 2016).

This de-regulation (or shift in regulation) of traditional industrial production-oriented forestry, and shift of focus to non-industrial benefits, is sometimes described as a transition to 'post-productivism' (Mather 2001). Instead of the industrial 'monofunctionality' of high-productivity, even-age tree plantations, in mainland Europe and North America there has been a change towards 'multifunctionality' with 'post-industrial forests' characterized by diverse landscapes and multiple uses, tended by ecosystem management (Veenman *et al.* 2009).

In recent decades, efforts to expand ideas on forest benefits have often been connected to the concept of ecosystem services, defined by the Millennium Ecosystem Assessment as 'the benefits people obtain from ecosystems' (MEA 2005). This includes provisional, regulative and cultural services, which directly influence human interests, and supporting services that assist delivery of the other three services (Costanza *et al.* 1997; Norgaard 2010).

Although the industrial use of forest assets in Sweden has faced pressure it is still very strong, in international terms (Lindkvist *et al.* 2011; Sandström and Sténs 2015). Recently the idea of bioeconomy has been promoted in international and national arenas (Staffas *et al.* 2013). The OECD (2006) defines this as 'transforming life science knowledge into new, sustainable, eco-efficient and competitive products'. It includes replacement of fossil-based raw materials with bio-based alternatives, development and implementation of innovative processes and new value chains, and a push to increase climate change mitigation (European Commission 2012; Formas 2012). In Sweden, the bioeconomy agenda has strongly influenced the forest arena (Pülzl *et al.* 2014). However, the term bioeconomy encompasses various 'shades of green', with different actors embracing different aspects of the concept (Kleinschmidt *et al.* 2014). Thus, although the bioeconomy agenda has been strongly focused on production and economic growth so far, there are efforts to integrate stronger environmental, sustainability, and public benefits components.

1.6 Concluding remarks

As shown in this chapter, major historical transformations can be discerned, from largely agricultural and proto-industrial use of Swedish forest resources in the mid-nineteenth century, through efficient industrial timber production 100 years later, to a balancing-act between industrial interests, environmental and other societal concerns in the present day (Beland Lindahl *et al.* 2015; Mårald and Westholm 2016). However, no single use or benefit has completely dominated the forest arena at any time during this period. Industrial timber benefits were traded-off against agricultural, proto-industrial, and energy benefits during the first 100 years until the mid-twentieth century, and subsequently against social and environmental benefits following development of the modern welfare system and urbanized society. Further, with growing awareness of the interconnectedness of global society and new challenges such as climate change, the interest in bioenergy and climate mitigation services from forests has strengthened.

The property rights and relations between different kinds of forest landowners are key elements of the Swedish forest arena. After sweeping changes in forest ownership in the nineteenth century, as forest companies grew, the situation stabilized in the early twentieth century. Since then, half of the forestland has been owned by several hundred thousand small-scale private forest owners. Thus, characteristics, views, and competences of these owners must be considered by anyone seeking to analyze, influence or transform forest ways of doing in Sweden. A further salient change is that 25 per cent of these owners no longer live near their forest property (SFA 2014).

A further complication is that forest owners have never enjoyed complete, exclusive rights to use their land. During the nineteenth century the peasants, who were collectively the major forest owners, had strong private property rights in relation to the national state's authority to steer the forests' use and management (Eliasson 2002). However, locally embedded usufructuary rights provided access to the forest for other individuals and groups for grazing, hunting, reindeer herding, and gathering fuel, berries, or materials for construction and handicrafts. During the twentieth century many of these rights were regulated through the Right of Public Access, the Reindeer Management Right, and hunting legislation. The multiple overlapping rights of access to resources and benefits in the same landscapes have been accommodated to some degree by a tradition of cooperation, but they have also resulted in various tugs of war between different actors (Sténs *et al.* 2016). Moreover, strong voices have always argued that the forests should be regarded as common goods that provide important benefits for the nation, the citizens, public welfare, the environment and/or the climate, thereby promoting more regulations and restrictions of private property rights.

Overall, the shifting balances between multiple benefits, the structure of forest ownership, parallel activities in the same landscape, and overlapping property and usufructuary rights have shaped complex relations and interactions between actors and interests. The resulting distribution of power constrains both property owners' autonomy to decide what to do with their forests, and the state's ability to govern and control the forest from above. Consequently, efforts to change the forest governance, management and practices – ways of doing – to enable sustainable development must involve coordination of numerous actors and initiatives from both above and below in a polycentric context.

2 Forest knowledge and management

2.1 Introduction

Forestry has been practiced for millennia, but scientific forestry has a far shorter history. In the eighteenth and nineteenth centuries forestry researchers developed techniques to manage forests as productive and controllable systems sustainably generating high long-term timber yields. In parallel, often collaborative, efforts, both silvicultural practitioners and natural scientists (botanists, entomologists, plant pathologists, hydrologists) contributed understanding of forest dynamics and their management (ways of knowing and doing). Because forests and forestry are important parts of the fabric of society and culture, social scientists and humanities have more recently contributed important knowledge regarding the history, social dimensions, and governance of forest and forestry

Forest sciences have both theoretical and practical elements, and (hence) involve both ways of knowing and doing. Academic research focuses on universal laws and generally applicable models. This has enabled knowledge of forests to progress beyond empirical relationships based on local practices and concrete experience to rational perspectives and increasingly robust theoretical frameworks (Harwood 2005; 2010). In contrast, the strong user-focus within forest sciences encourages researchers to pay strong attention to external goals and the potential benefits and applications of their research (Elzinga 1985; Kaiserfeldt 2013). Within applied and goal-oriented research there has been a strong tendency to develop optimum solutions and streamlining, leading to large-scale 'one-size-fits-all' forestry (Scott 1998). However, research influenced by close contact with practitioners and different localities, clearly demonstrate that, when theoretical laws and large-scale solutions are applied, the particular context with site-specific conditions, social circumstances and the practitioner's motivations must be considered.

This chapter focuses on the development of forest research and management approaches, spanning the 'ways of knowing and doing' spectrum and searches for both general large-scale solutions and specific adapted solutions. The chapter also explores transformations in the understanding of forests as natural resources, and in the meanings and contextual uses of key concepts.

The evolution of such notions is connected to both transtemporal thinking and the forest arena. In turn, changing understandings of the forest and its wider functions have played a key role in shaping common ways of thinking and the formation of forest social contracts.

2.2 The making of future forests

A key driver for the development of silviculture was the local and regional shortage of wood raw material. In the eighteenth and early-nineteenth centuries, economic and mathematical methods – *Cameral Wissenschaft* – provided the theoretical foundation for scientific forestry (Lowood 1990). In northern Europe, these developments were linked to the formation of the modern nation-state (Ford 2016), especially in the small German states (before the unification of Germany during the nineteenth century). These states had limited natural resources within their borders, so maintenance of their national forest assets was important (Warde 2006). Thus, silviculture emerged in a particular societal context with certain objectives. Planned felling and replanting over a fixed period of time were intended to increase national control over the balance between yield and regrowth. It was in this context that the concept of *nachhaltende Nutzung* – sustainable use – was coined, referring to scientific management of forest to obtain continuous, sustainable production cycles (Hölzl 2010; Warde 2011; Radkau 2012).

The German school of forestry had huge international influence, leading (among other things) to foundation of the Swedish Forestry Institute in Stockholm to educate foresters in 1828 (Eliasson 2002). It also formulated several fundamental doctrines of scientific forestry (Dargavel and Johann 2013):

The first doctrine, 'measurement', concerned the importance of developing methods to measure trees, stands, forests, and their growth. This led to the accumulation of quantitative information that enabled the scientific evaluation of forests and use of science to guide its governance and management.

The second doctrine, 'tending', stressed the importance of rational management and efforts to create and renew healthy forests by both regeneration after felling and the 'restoration' of woodlands ruined by agricultural misuse or excessive timber exploitation. Agriculture served as an exemplary model for the forestry equivalent 'silviculture', although ploughing, sowing or planting, topping, thinning, and harvesting generally followed one-year rotations in Swedish agriculture, while the sylvicultural equivalents took about a century.

The third doctrine, 'profiting', emphasized the rational economic foundation of forestry. The school's proponents often metaphorically described the forest as a 'bank'; something to invest in, administer and thus generate long-term interest and profits. This fostered a certain transtemporal perspective, describing the forests abstractly in terms of capital that could be calculated, managed, and enhanced over generations. A central research issue arising from this perspective was timing, particularly the most profitable time in the rotation period to realize the accumulated yield by cutting down the trees.

Finally, the fourth doctrine was 'regulating'. An ideal managed forest (*Normalwald*) should consist of homogeneous stands with a balanced aged-class distribution. This would facilitate measurements of the 'standing volume', 'growing stock', and (hence) the 'allowable cut' that could be regularly extracted without risking long-term sustainability (Dargavel and Johann 2013).

Scientific forestry has been portrayed as a paradigmatic example of 'high modernist ideology' (Scott 1998). That is, a strong belief in linear progress based on advances in scientific knowledge, satisfaction of human needs, and increased control over nature. By applying knowledge and 'rational' management, foresters made the forested landscape legible. Mapping and statistics provided aggregated and standardized information about a few aspects – a simplified overview of parts of the complex reality – while other parts were ignored. This notion implicitly embraces a confidence in technological fixes and the possibility of achieving a single optimum solution that balances production and sustainability. Such confidence often triggered large-scale, state-led endeavours to control and reshape the current order to forge a superior future. Accordingly, high modernism includes a distinct temporal understanding. As Scott writes: 'The past is an impediment, a history that must be transcended; the present is the platform for launching plans for a better future' (p. 95).

An obstacle to such transcendence, particularly in terms of the 'measurement' doctrine, was the lack of reliable forest statistics. Since the mid-nineteenth century several efforts had been made to begin addressing this deficiency in Sweden, Finland, and Norway by estimating the balance between yield and regrowth. In the early 1920s all three countries established regular, very ambitious National Forest Inventories (Tomppo *et al*. 2010; Fridman *et al*. 2014). In addition to surveys of forest volumes and timber supplies, these inventories collected information about the ages, sizes and tree species of stands in the national forests, and their silvicultural status (whether or not they had been, or needed, cutting, and whether they had been clear-cut or selectively cut) (Tomppo *et al*. 2010). The inventory approach shared the same rationale as the *Normalwald* ideal; that comprehensive, reliable information would enable control of all forest assets and implementation of scientific forest management that anticipates (and resolves) future dilemmas and grasps opportunities.

2.3 Translating scientific forestry into local practices

Although scientific forestry can be seen as a rationalistic, top-down and controlling activity, it included other aspects. In Germany in the second half of the nineteenth century, it became clear that both the theoretical forestry models and desired goals had to be adapted to different natural and societal contexts (Hölzl 2010). There was also awareness in Sweden of the importance of local conditions, including variations in soil, water, elevation, and climate (Mårald *et al.* 2016). Moreover, there was widespread understanding among foresters that forests have wider 'functions', in addition to timber production (Langston 1995; Johann and Dargavel 2013). Thus, there was a recognized need for forest managers to acknowledge the societal and environmental importance of well-functioning watersheds, wildlife, soils, and climate (subsequently regarded as 'ecosystem services').

The connection between forests and climate was debated particularly intensively in the nineteenth century. Researchers such as Alexander von Humboldt, Karl Fraas, Joakim Frederik Schouw and George Perkins Marsh, perhaps the most well-known today due to his seminal book *Man and Nature* (1864), all warned of irreversible changes due to deforestation causing soil erosion, water scarcity, and a drier climate (Olwig 1980; Lowenthal 2000; Mårald 2002; Hall 2005). To counteract such negative trends, afforestation campaigns started, for example, in the Alps, the Mediterranean region, and the USA (Langston 1995; Gardner 2009; Dargavel and Johann 2013; Higgs 2014; Ford 2016). In Sweden, the heavily deforested county of Halland in the south was described as an 'African stone desert', creating a bad regional climate, which should be amended by planting trees to improve the conditions for surrounding agriculture and societies (*Underdånigt betänkande* 1856).

The translation of universal natural laws into practical measures adapted to local conditions required trials in the field. Thus, in 1902 the Swedish State Forest Research Institute was established to conduct practical field trials. To enable large-scale, systematic, and long-term experiments the institute established three experimental field stations in the early 1920s, one in the south, one in the middle, and one in the north of the country, thereby covering wide ranges of climatic and site conditions (Bendz 2011). Moreover, in 1915 the Forestry Institute was converted into The Royal College of Forestry, with the same location and board as the State Forest Research Institute. The higher education in forestry it provided included courses in silviculture, natural history, botany, geology, climate, hydrology, and chemistry. The education also included a lot of practice, excursions, and field-located learning for students to acquire practical knowledge and other ways of knowing and doing.

However, for several reasons the Forest Research Institute and Royal College of Forestry became a stronghold of 'pioneer or proto-ecologists' (Söderqvist 1986). No doctorates in forestry were conferred in Sweden until 1950, so many of the first forest researchers were recruited from natural science departments. Moreover, in addition to forestry, some research at the Forest Research Institute focused on the natural history of forests, invasions of tree species, soil and water conditions, the evolution of plant communities, nature conservation issues, and wildlife management. Nevertheless, the main

Figure 2.1 Assistant Henrik Hesselman by the 'Jägmästarkojan' (the foresters'
cabin) conducting field trials in Hamra kronopark, Darlecarlia 1903.
Hesselman was a botanist and in 1912 became professor and head of
the Forest Research Institute.

Photo: Gunnar Andersson and Henrik Hesselman. Copyright Swedish University of Agricultural
Sciences.

priority remained ecological understanding that would facilitate establishment of rational forest management and improve timber production (Mårald
et al. 2016).

At the turn of the nineteenth century in Sweden the most intensively
debated research topic was the optimal forest management system, particularly clear-cutting or selective felling (Öckerman 1996; Lisberg Jensen
2011). Among foresters there was a general consensus about the overarching
goal – rational forestry that generated long-term sustainable yields to support
the associated industries and national economic growth – but the optimal
way to achieve it was debated. Selective felling dominated in the nineteenth
century, but the German approaches and industrialization inspired the first
introduction of clear-cutting and artificial regeneration systems (Lundmark
et al. 2013). In 1897, a leading Swedish forester, Uno Wallmo, published a

book called *Rationell skogsafverkning* (Rational felling) strongly advocating selective felling. The book provoked intense discussion, illustrating the conviction that there is a single optimum solution – a best practice – that, with some adaptations to local conditions, could be applied everywhere.

2.4 Optimization for timber production

After World War II, forest researchers became more practically oriented and involved in large-scale management schemes. As described in the previous chapter, this period was characterized by intensive timber harvest, triggered by economic growth and increases in construction activity. For Sweden, forestry was vital to enhance economic growth and social progress. As Thorsten Streyffert, vice-chancellor of the Royal College of Forestry and Professor of Forest Economy, claimed, 'if research in the production area is to improve our standard of living, this can also be expressed as a means to help us to produce more goods, better goods, and cheaper goods' (Streyffert 1961). Consequently, the aim for forest research was to optimize high timber production. This view of the importance of forestry was not only embraced in Sweden; the US Forest Service also expressed strong trust in progressive development governed by science and driven by new technology (Hirt 1994).

However, these desires sharply contrasted with empirically determined states of the forests. The results of the first national forest inventories, which finished in the 1930s, raised major concerns about the silvicultural quality of forests, especially in the northern region. A common opinion among forest researchers was that forests were composed of 'trash stands', 'green lies', and 'residual stands' – regarded as degraded due to agricultural misuse and a century of commercial exploitation through selective felling (Linder and Östlund 1998; Öckerman 1998; Lisberg Jensen 2011; Lundmark *et al.* 2013). Moreover, the forest inventories predicted that the poor forest conditions and growing demand for timber would lead to a 'timber famine', threatening the forest industry and thus national welfare (Mårald and Westholm 2016). Hence, this linear temporal notion implied that the extrapolation of the adverse historical development into a future dilemma clearly indicated that something had to be done in the present.

In this context, the foresters' idea of 'restoration' became salient. In 1950, the Director General of the Swedish National Forest Enterprise, Erik Höjer, launched a large-scale restoration programme for the vast state-owned forests in the north (Cirkulärskrivelse 1950). The remaining traces of agricultural forest use and selective felling were to be erased, and forests replanted to create even-aged productive stands. This intensive forest management system, which was also adopted by forest companies, included not only the clear-cutting of large tracts, but also slash burning or radical scarification of the ground, wetland drainage, replanting, and thinning. These silvicultural changes were accompanied, and partly driven by, the mechan-

Figure 2.2 A view of a clear-cut forest landscape. Forest excursion in 1976.
Photo: Evert Jeansson. Copyright Swedish University of Agricultural Sciences.

ization of forestry and standardization of best practice. Later, in the 1960s, fertilizers, defoliants and to some extent exotic tree species were used to increase production (Lindkvist *et al.* 2011; Siiskonen 2013; Backman and Mårald 2016). Overall, the restoration was described as a forcible 'investment in the future', which was expected to yield dividends many decades later in the form of considerable increases in forest volume (Mårald and Westholm 2016).

It should be noted that the quality of the forest research was not necessarily impaired by the strong focus on implementing a specific way of doing. On the contrary, with clear societal goals, strong state support, and close industrial contacts, Swedish forest research became world-leading, notably in tree breeding and forestry planning. For example, Swedish researchers developed procedures for producing tree seeds at operational scales that were adopted globally. This involved collecting seeds from 'plus trees' in the forest from different provenances, establishing seed orchards, controlled cross-pollinations between orchard trees, and progeny trials to assess and select the best performing parent trees. The seeds from the remaining parent trees could then be used to raise plants in nurseries for regular plantations (Dargavel and Johann 2013). In addition, advanced statistical methods were developed to handle the large amounts of data generated by the exhaustive national forest inventory and breeding programs (Matérn 1960; Dargavel and Johann 2013). In a larger context, national-level forestry planning was introduced to coordinate state forestry, and

Figure 2.3 National forest inventory in 1966. In the background professor in
        forest statistics Bertil Matérn guiding visiting foresters from India and
        Venezuela.

Photo: Jaroslaw Jaremko. Copyright Swedish University of Agricultural Sciences.

encourage forest companies and the numerous small-scale forest owners to
implement identified best practices.

Despite the primary focus on optimizing timber production, there was an
awareness that forests are complex biophysical systems that require multiple
scientific approaches to analyze. This is illustrated by the establishment,
in 1960, of a chair in forest ecology at the Royal College of Forestry – the
first explicitly 'ecology'-related professorship in Sweden (Söderqvist 1986).
As in the earlier 'proto-ecology' phase, forest research during the post-war
period mainly focused on nutrient-related aspects of forest growth with the
ultimate aim to enhance forestry regimes. However, in the 1960s and 1970s
ecology gained ground as a theoretical science (Bowler 1992). Eugene and
Howard Odum's notion of 'systems ecology', emphasizing energy flows
and feedback loops in ecosystems, became especially influential. Inspired by
economic models, such analysis charted 'energy budgets' and 'nutrient
capital', as well as seeking to establish the 'productivity', 'yield', and
'efficiency' of focal ecosystems (Worster 1994). Consequently, the system
ecology approach was close to the silvicultural doctrine of 'profiting'. As the
environmental historian Donald Worster noted: 'As a modernized economic
system, nature now becomes a corporate state, a chain of factories, an
assembly line' (Worster 1994).

Other influences included the International Biological Programme (IBP,
1964–1974), which investigated the productivity of biological resources and

human welfare (Söderqvist 1986; Coleman 2010), and the American Hubbard Brook ecosystem study, exploring a whole watershed of a forested ecosystem (Bormann and Likens 1979). Following such international ecosystem projects, the large-scale research programme Ecology of the Swedish Coniferous Forest Landscape started in the early 1970s involving several universities, one of which was the Royal College of Forestry (Söderqvist 1986).

2.5 The development of alternative forest management approaches

As outlined in the previous chapter, environmental awakening challenged the strong link between the forest industry and forest research, initiating changes in traditional forestry research and education. In 1977 a new 'sector university' was formed, the Swedish University of Agricultural Sciences, through an amalgamation of the agricultural, forest and veterinary colleges and research institutions. Similarly, the university's activities included education and research about land use, the environment and development assistance, and basic research in life sciences (Bruno 2016). Hence, this entailed a clear shift towards academic research and ways of knowing, and the new university became more like other universities, but still with a special responsibility for the 'green sector' (Mårald 2011).

There were also social links in other directions, connecting the forest arena to nature conservationism, environmental movements, and both national and international policy developments in the emergent environmental sphere (Warde and Sörlin 2015). Moreover, the need for a balance between economic development and conservation of the environment was formally recognized, both in Sweden and internationally, notably at the first global UN Conference on the Human Environment in Stockholm in 1972.

In the mid-1980s the concept of biodiversity became prominent. It is often claimed that the word was first used at an American nature conservation conference in 1986 (Takacs 1996), although the complete term 'biological diversity' had been sporadically used before 1980 (Farnham 2007). Moreover, Charles Darwin's *On the Origin of Species* and theory of evolution (Darwin 1859) had highlighted connections between places and the diversity of species more than a century earlier (Bowler 1992), and Alfred Russel Wallace (who independently conceived the theory of evolution through natural selection) warned against the 'extinction of the numerous forms of life which the progress of cultivation invariably entails' (Wallace 1863). However, the extensive use of the concept since the mid-1980s shows that it met a long-felt need (Figure 1.2). In related efforts, the International Union for Conservation of Nature (IUCN), founded in 1948 and supported by the UN, started a body called the Survival Service Commission in the 1960s to compile 'red lists' of species threatened by extinction (Gustafsson and Lidskog 2013). Continuing work since then has included efforts to

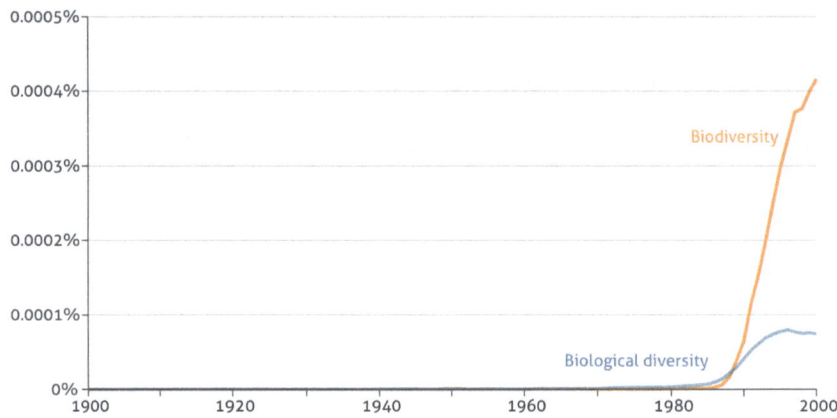

Figure 2.4 The skyrocketing use of the biodiversity concept. The term was rarely
mentioned before the mid-1980s. Just seven years later it was in the
limelight at the Earth Summit in Rio de Janeiro in 1992, which lay the
foundations of the Convention of Biological Diversity.

Source: Google Books' Ngram Viewer. Graphical design: Jerker Lokrantz/Azote.

standardize assessment procedures when compiling these lists and establish
a network of national partners.

The term 'biodiversity' bridged a gap between natural science theories,
nature conservation, and politics – thus between different ways of knowing
and doing. A major advantage was that it provided ways for measuring
damage to natural systems, which were applied particularly intensively in
Sweden and Finland (Berglund 2000; Gustafsson and Lidskog 2013).
Counting and mapping red-listed species, which were claimed to indicate
levels of threats for whole habitats, quantified complex ecological changes
(Turnhout *et al.* 2013), providing 'portable representations' that could be
readily analyzed and compared at scales ranging from local to global (Latour
1987; Lidskog 2014). Moreover, this made nature governable, as it enabled
politicians to establish goals, give directions for actions, and assess the
actions' impact (Asdal 2011).

Biodiversity was considered a resource that supported (while 'loss of
biodiversity' jeopardized) development and human welfare (Lisberg Jensen
2002). In this context, intensive forestry was assumed to weaken sustainability and the ability to meet future challenges (Bush 2010). Seeing biodiversity as a fundamental component of sustainable development, participants
in the UN Environment Programme conceived the idea of a Convention on
Biological Diversity in the late 1980s, which was drafted and finally signed
at the UN's 'Earth Summit' in Rio de Janeiro in 1992.

This political development also boosted new forms of integrated natural
resource management frameworks, including the Ecosystem Approach,
Ecosystem-based Management, and Resilience Thinking (all of which are

described and discussed in Chapter 7). Originating from natural science, these approaches crossed-over to the societal sphere, thereby including humans and social and economic goals as key components, within a wider eco-system framework (Christensen *et al.* 1996). With a normative basis, these approaches were intended to sustain ecosystems and reduce their vulner-ability, thereby safeguarding their capacity to meet current and future ecological and human needs. In this context, adaptive management developed as a method of stewardship for maintaining (or enhancing) systems' adaptive capacity and avoiding dramatic regime shifts (Holling 1973; Folke *et al.* 2002). In contrast to the optimization thinking that has dominated production-oriented forest research, adaptive management focuses on flexibility and diversity, recognizing the need to cope with unexpected changes that may occur.

From the 1980s onwards, social sciences increasingly contributed to forest research, focusing on forests and forestry as parts of the social fabric – involving webs of power relations, societal arrangements, human behaviours and cultural worldviews. From this perspective, environmental problems can be regarded as societal problems too, rather than external issues with no socio-economic implications (Beck 1992). Thus, a key task is to understand and guide social transformations. Recognition of this has fostered the emergence of integrated approaches, such as the Social-Ecological System Framework and Reflexive Governance (see Chapter 7). These approaches are critical of one-size-fits-all methods and end-solutions (Ostrom and Cox 2010; Voß and Bornemann 2011), and recognize societies as both inherent parts of the problems, and means to handle them. It is therefore important to understand how resources are embedded in polycentric social-ecological systems (Ostrom 2011), and encourage a reflexive understanding of governance and competing framings of the problems (Voß and Bornemann 2011).

All these approaches, predominantly rooted in either natural or social sciences, may involve social dimensions and participatory processes with concerned parties. In this context transdisciplinary, participatory, and social learning processes have been developed to identify and discuss uncertainties, conflicting goals, and divergent benefits to enable common action (Muro and Jeffrey 2008; Rodela 2013; Hicks *et al.* 2016). In the Swedish forest arena, with many small-scale private forest owners, it is especially demanding to translate and embed forest ecosystem management and climate mitigation measures. Forest practitioners are individuals working in different con-texts, with different ways of knowing and doing, with different experiences and different goals guiding their activities (Uggla and Lidskog 2015). The forest owners must be selective, and carefully consider the knowledge and management actions that are relevant and meaningful for their own forest practice (Lidskog and Löfmarck 2015; 2016). Hence, top-down communi-cation must be balanced with situated understandings and bottom-up processes to meet forest owners' needs (Eriksson 2015; 2017; Wallin 2017).

2.6 Concluding remarks

In the mid-1800s, the introduction of silvicultural research played a key role in the establishment and professionalization of modern forestry. However, although inspired by grand, generalized management theories and ways of knowing, the research was still infinitesimal and forestry management and practices – ways of doing – were very diverse. In the early twentieth century silvicultural research had been institutionalized, connecting forestry to academic research, especially natural sciences highlighting the importance of field trials to understand specific biophysical site conditions for forestry. Nevertheless, throughout this period silvicultural research was largely focused on optimization, to establish best practices in forest management to achieve certain goals, particularly high levels of production. In the post-war era it finally became possible to accomplish this goal. With close contact with the industry, strong support from the state and mechanization, silvicultural research focused on implementing the same management system across the whole country.

In the 1960s and 1970s the environmental awakening and expansion of environmental research challenged the narrow focus on production and established alternative ways of knowing and doing. Researchers broadened their perspectives to consider complexity, feedback loops, and dynamics, and developed methods to reduce uncertainty and enhance resilience to forest threats. The changes included greater efforts to generalize findings and establish theories. However, other kinds of social goal-orientation evolved, connected to the development of national and international environmental policy through the application of integrated approaches to explore both the social and economic dimensions of the problems. Since these approaches emanate from the intersection between global sustainability debates and general natural scientific perspectives, they have often been difficult to translate into concrete management programmes involving local concerned parties. In response, environmental researchers, social scientists, and humanists have elaborated collaborative and participatory processes to adapt management not only to different ecological conditions, but also to different social settings, societal scales, cultures and people's motivations.

3   Forest governance

3.1 Introduction

Organising collective action has been a major challenge for governing bodies in Sweden, partly because it requires the engagement of both large forest companies and hundreds of thousands of small-scale private forest owners. Moreover, international developments and other national challenges have raised further governance problems. Until the second half of the twentieth century the political regulation of forestland was mainly a domestic matter, under the aegis of the national government, despite influences from international forest-related debates. Since then, forest politics and governance have become much more globalized, influenced by transnational organizations, debates, and international politics. Today national forest politics is dictated and restricted by international law, including several conventions that are not primarily intended to regulate forests or forestry. Nevertheless, the implementation of international law follows diverse national paths adapted to different conditions, interests and traditions of ways of doing (Lindstad and Solberg 2012; Backman and Mårald 2016).

The forest industry, and the forest as a societal arena, have persisted and prospered in the face of changing national and international conditions. This has been partly due to the ability to establish stable, underlying agreements between the state, forest companies, forest owners, and the rest of society. Such forest social contracts are sets of implicit or explicit arrangements built on relational rights and duties between involved stakeholders to enable collective action in certain directions. However, such contracts do not involve all concerned parties or all relevant benefits and understandings. Instead, a forest social contract always involves delimitation of those seen as stakeholders, what are described as relevant benefits, and the understandings regarded as rational. Thus, a certain forest social contract reflects certain power relations.

This chapter focuses on the establishment and destabilization of three consecutive forest social contracts, reflecting the transformations of forest benefits, rights, and ways of knowing and doing described in previous chapters. It highlights how the main challenges addressed by each contract were framed, who were considered the concerned stakeholders, how competing demands were handled, and central methods of ways of doing. With changing conditions, caused by international and national develop-ments, these contracts have gradually lost their relevance and legitimacy, and in each case a new societal agreement has been negotiated, involving new understandings, actors and governing models.

3.2 The first social contract: governing industrial forestry

Rapid forest exploitation during the second half of the nineteenth century raised questions about private ownership, the responsibilities of large forest companies, and the role of the state. During the first half of the century liberal forest legislation prevailed, with power mainly distributed at the local level among many farming forest owners (Eliasson 1997; 2002). The entrance of large forest companies in the forest arena profoundly changed the conditions, as described in Chapter 2, especially in the northern region. The primary motive of the companies was to make profits. From a national governmental perspective this was short-sighted behaviour, threatening local communities as well as the national forest assets. However, these companies represented modern forestry and hopes for an industrial future for Sweden, which regarded itself as lagging behind the rest of Europe. Thus, the governmental bodies tried to simultaneously mitigate negative effects of the societal transformation and propel industrialization and a more production-oriented use of private forests. This balancing-act of encouraging and controlling industrialization was the challenge that led to establishment of the first social contract.

Thus, industrialization of forests in the mid-nineteenth century was accompanied by an ambition to govern the forest arena centrally. In 1855, the Swedish government appointed a Forest Committee to investigate whether it was possible 'to anticipate the devastation of privately owned forests and if the state had the legal obligation and duty to superintend' them (*Underdånigt betänkande 1856*). The Committee's report called for increased governmental capacity to govern forests with stronger legislation and establishment of a central authority with sufficient resources to conduct surveys of local conditions and play an educational role. Many of these recommendations were not fulfilled at the time due to a lack of political support, but in 1859 the National Forestry Board was established and assigned responsibility for state-owned forests and forestry education (Eliasson 2002).

A second Forest Committee was appointed in Sweden in 1896, to address concerns regarding industrial forestry and deforestation. The key problem was not logging *per se*, since a steady supply of timber was essential for the industry and its economic contribution, but rather the generally ignored long-term perspective and need for regeneration after logging. Thus, the

Figure 3.1 The sowing of new forests required a lot of work. Often whole families or school children were involved. Jörn, Västerbotten 1917.

Photo: Edvard Wibeck. Copyright Swedish University of Agricultural Sciences.

Committee sought to ensure sustainable yields by recommending imple-mentation of a forest management regime that aimed to provide high periodic timber supplies without impairing the sustainability of forest pro-ductivity. As a law regulating privately owned forests, the resulting Forest Conservation Act of 1903 signified the reinforcement of top-down legal governance. The most important regulation was a requirement for forest regeneration after felling (Stjernquist 1973). Moreover, County Forestry Boards were established to promote forest regeneration by disseminating information, distributing financial aid, and providing seeds and saplings to forest owners. Awareness regarding silviculture measures and long-term investments in the forest was supposed to be increased largely through cooperation, dialogue and persuasion (Appelstrand 2007).

The challenges posed by forest exploitation debated at this time included not only the threat it posed to long-term forest viability, but also the poverty and social misery it caused in the northern provinces (Sörlin 1988). The questionable means large forest companies used to purchase forestland from minor forest owners were particular concerns, as the deals often left the smallholders with insufficient resources to support their families. Consequently, the Swedish government prohibited companies from buying more forest in northern Sweden in 1906. Similar sharp dilemmas in the other

Nordic countries led (among other things) to prohibitions of the establishment of any new private forest companies in Norway (Eckerberg 1995), and company purchases in Finland (Siiskonen 2013).

Overall, an underlying societal agreement between the state, small-scale forest owners and forest industry can be discerned. The main challenges were framed in such a way as to support industrialization while securing the financial viability of farmers as small-scale forest owners and long-term sustainable supplies of forest resources. Under the agreement farmers collectively owned the largest share of forest, and the forest industry had to accept this. However, the small-scale forest owners had to accept that government-appointed professionals supervized and controlled forest management, to increase production to meet requirements of the forest industry. Hence, forest companies could continue to expand without obtaining more forestland because they were supplied with material from all forest owners nationally. The state increased its power and influence over private forests but wielded power mainly via 'soft law', through collaboration, economic incentives, advice, education, and persuasion (Applestrand 2007). Due to these changes forests and forestry was transformed, from a largely locally regulated 'appendage' of agriculture and proto-industry into a national industrial endeavour with all stakeholders sharing the economic risks and benefits (Donner-Amnell 2004; Lehtinen *et al.* 2004).

3.3 The second contract: forestry for national welfare

The post-World War II era brought economic growth, urbanization, modernization, and the expansion of welfare institutions. In Scandinavia, this expansion included dramatic development of the 'Nordic Welfare Model', based on the idea that the state should redistribute wealth, guide the building of a modern society and spread prosperity to every citizen in the nation (Hilson 2008; Árnason and Wittrock 2012). The efficient use of natural resources was an essential part of this modern project. The main challenges in this respect were that forestry practices were poorly developed, and undertaken by actors who were still only partially integrated in the market economy and still relied on simple technology. Moreover, industries offering better salaries and working conditions absorbed workers, making it increasingly difficult to maintain the forestry workforce. Thus, the low productivity in forestry threatened to reduce the forest industry's international competitiveness, thereby endangering the whole welfare project.

The answer to this was to rationalize forestry work, streamline the forest assets, and improve conditions for everyone involved in the forest industry. Reciprocally, they were all expected to contribute to efforts to create a strong and prosperous forest industry. In other words, while the first social contract can be described as reactive, a way to gain control over industrialization and alleviate its negative consequences, the second contract was formed proactively, and offered a way to shape a better future.

Consequently, after World War II a further step in state intervention was taken. The Swedish Forest Conservation Act of 1948 expressed a belief that governance needed a strong institutional base and a top-down approach. It reinforced the National Forestry Board and established a network of national, regional, and local foresters to monitor and implement modern forestry (Appelstrand 2007; Enander 2007). The strong governance of privately owned forests and forestry was not solely based on legislation; there was also an overarching consensus between leading politicians, forest companies, and trade unions regarding the goals and means of forestry. Increases in timber production and the industry's efficiency would provide returns on investments in modern forestry with interest for the forest owners, and strengthen the forest companies by making their long-term capital investments foreseeable. In turn this would contribute to national economic growth and employment, ultimately generating public welfare for all citizens.

Thus, compared to the earlier 'contract' there was more focus on welfare and development for the whole country. With this wide agreement as a legitimate foundation, another logical step was state provision of subsidies to support private forest owners (both small-scale and large forest companies). Hence, subsidies were provided for regeneration, forest ditching, building forest truck roads, and 'restoration' of low-productivity forest stands. There were also special subsidies for forestry in sparsely populated areas in northern Sweden (Bernes and Lundgren 2009). By the same token,

Figure 3.2 Mechanized forestry 1962. Svartliden, Sorsele, Lapland.
Photo: Helge Granbom. Copyright Swedish University of Agricultural Sciences.

the state took responsibility for bolstering forest research, education, advisory services and infrastructure – for example, the reinforcement of road bridges that were important for transporting timber and mechanized forestry. Without these subsidies the implementation of modern clear-cutting forestry would probably have been much less economically feasible.

The gradual increase in regulation, state intervention and promotion of the forest industry, which had lasted for more than 100 years, peaked around 1980. The objective of the Swedish Forest Conservation Act of 1979 (with some amendments in 1981 and 1983) was to ensure a steady supply of timber for the forest industry by prioritizing forest regeneration, the thinning of young forests, and the replanting of unproductive stands. In addition, selective cutting was expressly forbidden (for the first time, although it had ceased in practice since 1948). Moreover, it became compulsory for private forest owners to have forest management plans, and to fell some parts of their old-growth forests (Enander 2007; 2011). This was the first time that Swedish forest regulation shifted from 'soft' to 'hard law'; that is, towards enforcement (Appelstrand 2012).

The rationale underlying the second social contract was that increases in production would enable economic growth and progressive societal trans-formation. Intensive forestry was legitimized by the perceived possibility it offered to expand welfare to everyone. Involved stakeholders were the state, the forest industry, forest owners, trade unions, and (according to the rhetoric) ultimately all citizens of the welfare state. Consequently, the forest owners lost more of their independence regarding management goals and approaches. Their duty was to manage their forest to produce timber and supply the forest industry. In compensation, farmers, lumberjacks, and small-scale forest owners acquired access to welfare services, and the state subsidized forest assessment to increase production. Power was mainly exercized collaboratively between the dominant actors, but also through 'hard law' to enable the ways of doing to ensure that every forest owner followed recommended practices. Hence, this contract had a strong emphasis on collective action; everyone should manage their forests in the same way to meet the same common goal.

3.4 The third social contract: the Nordic Forestry Model

In the 1990s, the Swedish forest arena sharply shifted due to globalization, changes in political pressures regarding the balance between forest production and environmental interests, and new approaches to governance. This new societal agreement is usually described as 'the Swedish Forestry Model' (KSLA 2009; Sandström and Sténs 2015; Beland Lindahl *et al.* 2016), or 'the Nordic Forestry Model', as it was also adopted in Finland (Donner-Amnell 2004; Lehtinien 2008; Beland Lindahl *et al.* 2015). In general terms, this model is associated with a combination of competitive, high productivity forestry and increases in forested areas. Moreover, it is described as an accessible way to maintain biodiversity and biological structures in land-scapes focused on forest production through joint efforts by the state, forestry industry, and forest owners. It, in the way it has been named, also alludes to the Nordic welfare model, with a long tradition of seeking consensus and balancing multiple interests for the common good.

This shift was preceded by a more strident environmental critique of modern forestry, strengthening of environmental organizations, and wide acceptance of the biodiversity concept and concerns. Structural rationalization in the forest industry also reduced numbers of employees in the sector. Several closures of saw and paper mills, and other production units, raised public concerns and showed that the forest industry was no longer delivering national welfare everywhere and to everyone (Donner-Amnell 2004). Thus, the foundation of the second contract eroded.

Moreover, in this context the 'sustainable development' concept entered the political stage. Compared to the old forestry concept of sustainable yield, this term had a broader meaning, developed by the UN's World Commission on Environment and Development (WCED) led by the former Norwegian prime minister Gro Harlem Brundtland. The report of the Brundtland Commission, *Our Common Future* (1987), defined sustainable development as 'development that meets the needs of the present without compromising the ability of future generations to meet their own needs.' By stressing that economic growth, environmental sustainability, and social equity are mutually interdependent, the report showed a way forward that could bring into play different actors and interests globally (Borowy 2015). Sustainable development became the leading principle of the UN's action plan Agenda 21, established at the 'Earth Summit' in Rio de Janeiro 1992, which subsequently consolidated policy-making and ways of doing on all levels.

National environmental organizations and local environmental groups started to connect and create international networks. Of most relevance to the Swedish situation were Canada and northwestern USA, and in all these regions environmental organizations had protested against intensive forestry since the 1960s. With limited success at the national level, domestic groups turned to transnational networks to pursue their goals (Keck and Sikkink 2014). Links to German environmental organizations were especially important, as Germany was an important wood commodity market (Soyez 2000). With headlines like '*Kahlschlag in Paradise*' (clear-cutting in Paradise) in the influential magazine *Die Spiegel*, campaigns for 'clear-cut-free paper', and threats of boycotts by the large Springer Verlag publishing company, pressure was imposed on national forest policy and conventional forestry practices (Berglund 2001; Baldwin 2003; Bush 2010; Winkel 2014; Simonsson 2016).

Such global connections inspired Swedish environmental organizations in the early 1990s. At the Earth Summit in Rio de Janeiro the Swedish Society for Nature Conservation condemned Swedish forestry for causing a severe loss of biodiversity, generating significant international and domestic attention. In 1992 in northern Sweden, inspired by tropical forest networks, the Taiga Rescue Network was started to connect environmental groups in the global North with a focus on the boreal forest zone. One strategy was to focus international support on specific 'hotspots' – that is, biologic-ally valuable forest areas threatened by logging – to overcome forest industries' and forest owners' resistance to adopting more biodiversity-promoting practices (Hellström and Rytilä 1998; Lisberg Jensen 2002). While forest industry capital, in the form of manufacturing infrastructure and forest properties, can be immobile and locked in specific places, inter-national environmentalism can be highly mobile (Hayter and Soyez 1996). Consequently, the enhanced global collaboration and norm creation drew the national consensus in forest politics away from the previous prioritization of production towards biodiversity.

The altered balance between production and environmental goals was manifested in the Swedish Forestry Act of 1993, the first paragraph of which states: 'The forest is a national resource. It shall be managed in such a way as to provide a valuable yield and at the same time preserve biodiversity' (Skogsvårdslag 1993). While these two objectives are treated as 'equal', the Act also states that social, cultural, and aesthetic values, and Sami reindeer herding, should be cared for. In addition, the sixteen National Environmen-tal Quality Objectives, established in 1999 and 2005 (Sweden's environ-mental objectives 2006), shifted priorities in the forest sector from sustainable production to handling multiple goals.

The governance of forest also changed radically. Liberalization and deregulation of the forest sector followed a general international neoliberal turn in the 1980s and 1990s. In the environmental arena this is often described as an 'eco-modern' way to respond to environmental challenges (Hajer 1995). That is, an understanding that environmental goals and eco-nomic growth do not conflict – instead they go hand-in-hand, supporting each other. Via eco-innovations, eco-entrepreneurship, environmental label-ling of commercial products, the monetarization of environmental values and costs, establishment of markets and market solutions, and increasing consumers' awareness, the assumption was that it should be possible to create sustainable win-win situations that both benefit the environment and promote the growth of wealth.

This trend also strongly influenced international forest policy from the early 1990s onwards. Leading principles and ways to implement forest policy became: marketization, enhancement of the private sector's role, deregu-lation, and voluntarism (Humphreys 2006; 2014). The state's role became more a steering function through the release of market forces and activation of incentives for competition and development (Appelstrand 2012). This shift also involved a change of focus from 'government' to 'governance' (Kjær 2004; Arts and Buitzer 2009). While government refers to a formal body with the authority to make decisions in a given political system, governance concerns whole system management and decision-making processes by both formal bodies and market actors, civic society, organizations, and networks (Bevir 2013).

In this context 'freedom with responsibility' became the main Swedish principle underlying the relationship between the state and private forest owners. Thus, Swedish forest policy (which only slightly more than a decade earlier had introduced 'hard law') radically shifted back to 'soft law' with minimal requirements imposed by forest legislation (Applestrand 2012). As long as small-scale forest owners or forest companies stayed within the general limits of the law, they could decide how to achieve (and trade-off) the multiple forestry and environmental objectives. The forest owner became the main decision-maker, while the focus of national and regional forestry boards shifted from law enforcement to advising, monitoring, and evaluating. Moreover, to meet environmental targets in this deregulated environment, voluntary measures going beyond legal minimum levels were required. Consequently, a large proportion of Swedish forestland has been registered in the voluntary, market-based Forest Stewardship Council (FSC) and Programme for the Endorsement of Forest Certification (PEFC) certification schemes, developed in cooperation with various industrial, environmental, and social stakeholders (Johansson 2012; Johansson and Keskitalo 2014).

In practice, 'Retention Forestry' became the way to combine production and biodiversity targets for the same landscape. This refers to a modified form of clear-cutting that leaves part of each original stand unlogged to maintain the continuity of structural and compositional diversity – for example, by leaving forest cover along riparian zones, dead wood, and old trees (Gustafsson *et al.* 2012; Lindenmayer *et al.* 2012). As most biological diversity in Sweden was found in managed landscapes, it became important to incorporate production forests in the efforts to preserve biodiversity (Gustafsson *et al.* 2010). Even in the mid-1970s the Swedish Forestry Acts included amendments regarding considerations of nature in forest management (Simonsson *et al.* 2013; 2016).

Influences also came from abroad, notably modified forms of forest management developed in North America in response to the strong environmental criticism against clear-cutting forestry around 1990. One was the 'New Forestry' system, and its later refinement 'Ecosystem Management', introduced by the US Forest Service to meet environmental criticism (Freeman 2002). New Forestry focused on maintaining biological legacies and increasing spatial complexity. This influenced Swedish policy-making, and representatives from the Swedish Forest Agency visited north-western USA to see how the management system worked (Simonsson *et al.* 2013). From a societal perspective, both Retention Forestry and Ecosystem Management can be seen as results of a particular historical conflict situation and politically constructed compromises (Freeman 2002). Furthermore, Retention Forestry is 'a bottom-up conservation approach', since individual forest owners or forest managers are ultimately responsible for the performance and trade-offs between different goals in their forests (Gustafsson *et al.*

2012). In other words, it is no coincidence that Retention Forestry evolved contemporaneously with the idea of 'freedom with responsibility'.

The essence of the third social contract is the effort to balance real-time claims on the forests regarding production, environmental conservation, and (to some extent) other interests. This is done in a deregulated and neoliberally inspired context with market solutions and voluntarism guided by governmental steering. Important stakeholders are the forest industry, the environmental movement, other non-governmental organizations (NGOs), and market actors. The state still plays a vital role, but from a more withdrawn position, steering from the distance with soft law, goals, economic incentives, and monitoring. Instead, forest owners have been granted the right to manage the forest in accordance with their own goals, but with duties (mandatory and voluntary) to balance production and nature protection demands in practice. Moreover, the role of the Swedish population is not primarily as citizens in a welfare state, but as consumers in a market steering the forest sector by their active choices of certified and green-labelled wood commodities. Thus (it is hoped or presumed), rational and well-informed individual actions by forest owners, consumers, and market actors will result in collective action in desirable directions.

3.5 Concluding remarks

The three social contracts described and discussed in this chapter outline forest stakeholders' relative power and roles at the time they applied, and how common goals, mutual relations, trust, and long-term rules enabling common action were established. The main principle for implementing policies throughout the covered period has been 'soft law' – the provision of support, advice, persuasion and encouragement of cooperation – while 'hard law', enforcement, has been an exception.

The second contract can be seen as an evolution of the first, while the third constituted a much clearer shift involving environmental benefits and actors, as well as new ways of doing (Table 3.1). During the whole period the state, the forest owners, and the forest industry have been key actors, jointly defining the desired objective: a form of forestry that provides sustainable yields, efficiency, and profitability (Siiskonen 2013). In the last forty years other actors with other perspectives, most importantly environmental perspectives, have entered the forest arena. Moreover, some general shifts in forest government can be discerned. Until the late 1980s Old Public Administration – top-down, state-led 'rowing' towards a few politically defined objectives – dominated. The shift to the neoliberal Swedish Forestry Model entailed adoption of New Public Management, with policy-makers acting by 'steering' the market and fostering entrepreneurialism (Appelstrand 2012).

Despite tremendous changes in the social context of the Swedish forest arena during this period, old benefits, understandings, ways of knowing and

Table 3.1 Summary of the three Swedish forest contracts identified from c. 1900 to date

	1st contract c. 1900–1950	2nd contract c. 1950–1990	3rd contract c. 1990–
Framed challenge	Control forest exploitation and industrialization	Welfare for everyone	Balancing production and biodiversity
Direction of handling	Reactive	Proactive	Balancing real-time claims
Ways of governing	From liberalism to state intervention	Old Public Administration 'Rowing'	New Public Management 'Steering'
Policy implementation	Soft law Advice	Soft and hard law Advice Subsidies 'One-size-fits-all'	Soft law Goals 'Freedom with responsibility' Monitoring Market Voluntary action

doing, and 'contracts' do not tend to disappear when new ones emerge. Altogether this gives rise to path dependencies, legacies, and ways of thinking and doing that forge the present in a certain way.

Thus, a historical outlook forces us to widen the perspective and see the current situation and ourselves in a larger temporal context that gives the present depth it would otherwise lack. Historical outlooks likewise highlight transitions. Obviously much has changed during the course of history. This gives rise to the insight that the present situation, with our understandings, aspirations and ongoing struggles, is only transient, and so are we now living. Deep understanding of the present and awareness of the transience of the present highlight the need to leave a legacy, in terms of both tending the heritage of the past and making a lasting and beneficial impression on the future. Hence, this transtemporal thinking connects how historical layers and legacies morph into future trends, expectations, and desires. How this will affect the forest arena in Sweden is not set in stone. Such futures are the topic of the next part of this book – 'Looking forward'.

Part II
Looking forward

Revealing the future

The act of planting a tree is often seen as an investment in the future. A seedling today will develop into a mature tree that can provide a range of goods and services for future generations. However, choices of seeds and cultivation practices now will determine how trees grow and the benefits they will be able to provide. By planting specific trees and caring for them in certain ways, we are making choices for future generations. But what do we know about the future, and how does the future feed into the present?

Attempts to explore and control the future are as old as humanity and have taken many forms. Prophets, oracles, and seers drew on sources of privileged knowledge to access a future that was conceived as pre-existing destiny. Their efforts to unveil the future were based on the assumption that there already *is* a predetermined future to be discovered and told. Adams and Groves (2007) differentiate such efforts to learn what lies ahead from knowledge practices that intend to 'tame' the future; that is, to make daily life more predictable. The latter entails a quest for knowledge that can be used to structure, order, and handle the insecurities of the realm beyond experience. Rituals and religion, and all kinds of institutions, help make possible estimations and calculations about otherwise unpredictable events and interactions. Banks' efforts (since the Middle Ages) to calculate future monetary values in order to set interest and credit rates, insurance companies' calculations of future risks, and other initiatives to 'trade' the future are based on similar concerns (Adams and Groves 2007).

The need for governments to consider long-term developments, or 'the future', was increasingly recognized in the post-war period, due to the increasing rapidity of technological, social, and environmental change. Future studies as a field of science grew in response to needs for knowledge about likely future developments and for making informed choices and trade-offs between anticipated futures (Bell 1997; Kuosa 2011; Andersson 2012). Notably, concerns about the environment and natural resource management highlighted critical future issues (Warde and Sörlin 2015), which were incorporated into the sustainable development agenda (Robbin *et al.* 2013).

However, despite strong scientific inputs in debates, 'the future' is inherently political, connected to issues about choice, desires, power, and unequal possibilities to influence the development. Thus, the relation between science, politics, and the future is not unproblematic. In this respect, Andersson and Keizer (2014) have noted a 'democratic problem of the long term', rooted in the way predictive expertise is used when the future is addressed and incorporated into policy and knowledge systems. The key questions are: how should futures be represented, by whom and with what kind of balance between science, policy-making and public participation?

Hence, the study and representation of the future is not a purely scientific question. The future is open to alternative pathways and choices, so we must acknowledge and consider possible *futures* rather than a single set *future* (Masini 1993; Marien 2010). No future exists yet and, although there are many possible options for it, this does not mean that everybody can shape the future according to their preferences. Humans are constrained by natural processes beyond their control, by developments that are not foreseeable, by established societal structures, and by power relations that distribute influence unequally. There are also ongoing conflicts about definitions of future conditions, and establishing measurements in the present to anticipate them (Brown *et al.* 2000).

All future studies contribute to the construction of visions and formulation of statements that may influence future events, and thus participate in 'world making' (Sardar 2003; Adams and Grove 2007). Sardar (1993) critically describes this as 'colonization of the future', noting that authors of future studies often have agendas with purposes and goals that may or may not be explicitly stated (or even consciously recognized). There are also ongoing conflicts about definitions of future conditions, and establishing measurements in the present to anticipate them (Brown *et al.* 2000). This may in turn have real world consequences. For instance, notions of future timber shortages and increases in demand bolster strategies to promote timber production, which shape future forests in a highly material sense. Likewise, perceptions of future biodiversity losses may increase support for measures to protect habitats and ecological structures (Uggla *et al.* 2016). Today, threats posed by climate change are raised in all discussions about future forests, but in diverse ways and with diverse suggestions regarding the best ways to act to anticipate and address them (Winkel *et al.* 2011).

Consequently, there is a 'duality' in society's relation to the future. It is highly important to try to foresee coming challenges and possibilities in order to establish long-term plans that safeguard sustainable social, ecological, and economic development. Thus, it is essential to develop and apply robust methods to anticipate the future and establish strategies to cope with identified risks. On the other hand, future studies and 'world-making' always raise risks of colonising the future, which weakens voices of less powerful actors, and closes alternative pathways and opportunities.

According to the transtemporal thinking that this book promotes, the future and present are not linearly ordered and separate. Instead, the future is attached and constantly feeding into the present, so it is wise to increase openness to more, and richer, perceptions of the future, which may generate more creative pathways and identify ways to resolve present dilemmas. Likewise, transtemporal thinking increases awareness of the importance of responsible decision-making in the present, rather than postponing solutions of problems. This duality – the importance of future studies to enhance our capacity in the present and the risk of colonising the future – must be handled in future studies relevant to the forest arena, and for this reflexivity is essential.

In this part of the book we explore how the future can be scientifically approached, how future studies have been used in past and contemporary forest research, and characteristics of a reflexive way to handle the complexity of forest futures. The first chapter explores different methodologies for addressing and outlining the future. The next chapter focuses on how envisioning the future has changed over time in a Swedish forest context, and explores contemporary forest-oriented future studies. Finally, the last chapter discusses ways to handle the 'duality' by outlining a reflexive notion of future studies and discussing ways to develop forest future studies to obtain richer visions of the future that can feed into present decision-making and practical resolution of dilemmas.

4   Methods to study forests' futures

4.1 Introduction

A challenge for all futures studies is impossibility of obtaining empirical data, i.e. the future has not yet happened so it is genuinely unknown and uncertain. However, management decisions made in the present may affect forests for decades and even generations. Thus, the far-off future was traditionally regarded as 'dimly seen, veiled in the manager's uncertainty' (Duerr and Duerr 1975, p 31). Today, the futures of forests are increasingly uncertain, due to effects of multiple interacting factors, such as climate change, globalization, technical transformations, and associated policy and societal responses. Hence, various methodologies have been developed in environmental future-oriented research, and applied in both academic and forest management settings.

This chapter offers a brief overview of methods applied in environmental futures studies. In particular we focus on different forms of scenario analysis, and the first section describes a framework for scenario analysis. In its approximately fifty-year history, scenario analysis has developed into a methodology that, when correctly used, can display a multitude of possible futures and alternative pathways for future development. The next section discusses the use of quantitative and qualitative approaches in scenario analysis. Quantitative approaches dominated scenario analysis for a long time, but recently qualitative approaches have become more common. A current challenge in scenario analysis research is to find better methods to merge quantitative and qualitative analysis to combine the strengths of the two approaches. The last section deals with 'normative elements' in environmental future studies, i.e. desired ecological and/or socio-economic conditions (and the actions required to attain them), in contrast to non- or less normative contextual (descriptive) analyses (e.g. Milestad *et al.* 2014). Normative scenario analysis has increasingly recognized value in promoting social learning and change when current developments are regarded as undesirable or unsustainable. However, use of normative scenario analysis raises risks associated with the previously mentioned 'duality' of future studies, and requires caution to ensure that it is conducted and applied in a transparent, respectful, and reflexive manner. An alternative way to research how normative elements may shape the future is to explore stakeholders' norms, perceptions, and strategies for the future in interviews. Such studies must also consider the distribution of power between stakeholders, and thus their capacity to realize a future they perceive as desirable.

4.2 Scenario analysis facing the future

Scenario analysis is commonly used in future studies for envisioning a range of futures that might emanate from the complex dynamics of socio-ecological systems. Scenarios can be defined in many ways, but for more than a decade the Intergovernmental Panel for Climate Change (IPCC) has defined them as: 'images of the future, or alternative futures that are neither predictions nor forecasts, but an alternative image of how the future might unfold' (Nakicenovic and Swart 2000). Scenarios are increasingly used at local to global scales to explore the dynamics and sustainability of socio-ecological systems and to foster long-term thinking (O'Neill *et al.* 2017). Hence scenario analysis offers a tangible methodology for creative exchange between actors' different ways of knowing and doing.

Scenario development generally involves three main steps: 1) generation of ideas and gathering of data, 2) integration of the different parts into a viable framework, and 3) checking the developed scenarios' consistency and logics (Börjeson *et al.* 2006). The development (and subsequent analysis) approaches can be categorized as either intuitive or formal (van Notten *et al.* 2003). Intuitive processes lean strongly on qualitative knowledge, and typically involve creative techniques, such as the development of storylines. In contrast, formal approaches are based on quantitative knowledge and typically involve computer simulation techniques. Another important aspect of scenario development is the vantage point from which the scenario is developed. Exploratory or forecasting scenarios take the present as their starting point, while anticipatory or backcasting scenarios start from a specific future situation (van Notten *et al.* 2003). Hence, backcasting scenarios explore pathways that must be taken to arrive at a desirable future situation and thus are normative in nature (Dreborg 1996; Höjer and Mattsson 2000).

To structure the description and discussion of the various kinds of foresight and futures studies presented in this and the following chapters, we apply a common categorization of such studies: 1) 'What will happen?' (probable futures), 2) 'What could happen?' (possible futures), and 3) 'What should happen?' (preferable futures). As shown in Table 4.1, each of these categories can be subdivided into two sub-types that apply different approaches to address the questions defining the categories (Börjesson *et al.* 2006). 'Forecast' and 'What if?' types of probable futures studies respectively attempt to predict what will happen if a likely development occurs, or some specified event occurs. 'External' and 'Strategic' types of possible futures

Table 4.1 Summary of the three common categories, and sub-types, of futures studies in contemporary environmental foresight research (modified from Börjesson *et al.* 2006)

Question asked	Category	Sub-types
What will happen?	Probable future (predictive scenarios)	Forecast What if?
What could happen?	Possible futures (exploratory scenarios)	External Strategic
What should happen?	Preferable future (normative scenarios)	Preserving Transforming

studies respectively explore situations or developments that may happen if external factors we cannot control change, or we act in a specific way. Studies of preferable futures are inherently normative as they explore how a desirable target can be reached. 'Preserving' and 'Transforming' types respectively consider whether targets can be reached by adjusting the current situation or how prevailing structures must change. This is only one example of various typologies of scenario studies that have been developed during recent years; for others, see Hoogstra-Klein *et al.* (2016).

4.3 Balancing quantitative and qualitative knowledge

Dynamism and complexity are inherent properties of socio-ecological systems, and since the 1970s numerical models including numerous feedback loops have been developed in attempts to simulate human-environment interactions spanning many decades. Together with the availability of powerful computers, this has led to several large integrated models for analysing scenarios designed to explore global sustainability issues, which have informed many major scientifically based policy reports, e.g. the Club of Rome Report (Meadows *et al.* 1972), the Global Biodiversity Outlook 2010 (Pereira *et al.* 2010), the Global Energy Assessment (Johansson 2012), and the Intergovernmental Panel on Climate Change Assessments (IPCC 2013).

The most common type of integrated model used for scenario analysis in the forest sector considers forestry, forest industries, and their interactions. Such models have been extensively used in research since the 1980s and, although many models are used, they all have similarities in theoretical foundations and structure (Latta *et al.* 2013). Generally, they have been developed to analyze consequences for the forest sector of shifts in economic conditions and/or policy, i.e. predictive analysis of probable futures if some specified event occurs (What if? type scenarios; Table 4.1). However, there is increasing interest in strategic exploratory scenario analysis, as applied in the FAO's most recent European Forest Sector Outlook Study (United Nations 2011).

Forestry decision support systems (FDSSs), another type of models that plays an important role in the forest sector, consider mainly forests and the silvicultural practices that may be applied. Forestry planning inevitably includes projections of future forest development under possible silvicultural regimes. FDSSs are frequently used for such projections and can (for example) calculate sustainable harvest levels for a forest holding, region, or nation. A recent example is the scenario analysis of the Swedish forest for the upcoming 100 years by Claesson *et al.* (2015). Its reference scenario is based on a trend analysis of general practices in Swedish forestry during the last decade and estimates that forest growth will be approximately 20 per cent higher at the end of the 100-year simulation period than it is now due to climate change, if the management regime does not change much. In addition to the reference scenario, a number of alternative probable futures were modelled, focusing on long-term forest yield (Claesson *et al.* 2015). FDSSs are also increasingly used for projecting future effects of current silvicultural practices on forest ecosystem services other than tree growth. For example, Roberge *et al.* (2016) explored future forest biodiversity following application of different nature conservation strategies in the present, and Korosuo *et al.* (2014) explored effects of different silvicultural practices in the present on future reindeer husbandry.

Although quantitative scenario analysis was the most widely used technique for exploring key drivers of development for a long time, it is limited by the lack of connections to complex changes in social drivers, such as changing market demands, shifts in people's preferences, and/or policy developments. This is a major limitation because social, political, and economic shifts may have stronger effects than biophysical changes such as climatic changes, at least changes of the magnitude anticipated in Sweden during the coming centuries. Thus, it has been argued that creating rich and descriptive narratives may be as important as quantitative modelling, particularly in transdisciplinary scenario analysis, since narratives improve the scenarios' accessibility, credibility, and relevance (Burnam-Fink 2015). The development and dissemination of scenarios enriched with narratives also tend to involve strong interactions with diverse stakeholders, which can facilitate social learning. However, while applications of this qualitative and transdisciplinary approach in scenario analysis of future forests have been quite comprehensive in terms of identified drivers, to date they have involved no quantitative modelling (e.g. Moen *et al.* 2010, Pelli and den Herden 2013, Sandström *et al.* 2016).

Moreover, scenarios that combine qualitative and quantitative components have mostly considered national to global scale developments to date, although regional or local scenarios enriched with narratives connected to global and national developments may be particularly effective for engaging citizens and decision-makers in adaptive and transformational development processes. However, construction of such scenarios requires local and regional scenarios to be linked with quantitative modelling assessments spanning multiple spatial scales, and as yet there are few methods for such linkages, all of which require further development (Booth *et al.* 2016).

4.4 Normative futures and the influence of power relations

People living in the present have different ideas about preferable futures. Thus, as the future is socially constructed by people who have different understandings of (and visions for) the future, 'drivers of change' are embedded in the perceptions and strategies of those living in contemporary society (Beland Lindahl and Westholm 2012; Beland Lindahl 2015). However, different people have different capacities to transform their visions and statements into actions that shape future developments. So, indications of possible futures can be obtained by studying people's perceptions, strategies, power relations and interactions.

In the context of scenario analysis, norms and values are contentious because it can be justifiably argued that all scenarios are normative as they consist of the interpretations, values, and interests of the scenario developers. However, it is usually claimed that scenarios exploring probable or possible futures are more descriptive, and hence less normative, than scenarios that explore preferable futures (van Notten *et al.* 2003; Börjesson *et al.* 2006).

Backcasting is one kind of normative scenario approach that is particularly suitable when: the problem examined is complex and exacerbated by dominant trends; major, cross-sectoral changes are needed; and the time horizon is long enough to allow considerable scope for deliberate choice (i.e. Dreborg 1996). As already mentioned, there are always risks that dominant and powerful stakeholders will be able to define what the future will entail, while alternative trajectories will be largely ignored. This does not necessarily mean that the futures such stakeholders envision are optimal, or even realisable in practice. Rather, particular dominating narratives play a key role in justifying and constituting the visions and strategies that do come to be realized – and suppressing others, even if they are more technically and socially viable (Leach *et al.* 2010). In backcasting, stakeholders' *desired futures* are used as points of departure. Hence, as long as a broad range of stakeholders are involved, or considered, a broad range of preferred futures will be acknowledged and taken into consideration.

Backcasting is usually used to help efforts to understand not only how desired futures may vary between different stakeholders, but how desirable futures can be realized. Consequently, backcasting explicitly addresses the mechanisms whereby particular desired futures will materialize. The objective is to produce scenarios of a normative and transforming character to identify structures in the present system that must be overcome and changed, rather than preserved, to reach a desirable future (Börjeson *et al.* 2006; Dreborg 1996). This approach can be connected to Transitional Management, which will be described and discussed in the final part of this book.

4.5 Concluding remarks

This chapter discusses how future studies can be designed to foster prepared-ness for unknown and uncertain futures by opening up new avenues of thinking. The diverse methods available to study the future all have pros and cons, but many have complementary strengths. In forest research, quantitative modelling has unique capacities simultaneously to assess conse-quences of changes in multiple parameters across multiple temporal and geographical scales, and to handle enormous amounts of data. It provides critical knowledge about future possibilities by quantifying consequences and key relationships. Qualitative analysis builds on the creation of rich narratives of alternative forest futures, often with inputs from various stake-holders. Mixing quantitative and qualitative methods presents a method-ological challenge that few future studies in forest research have convincingly surmounted. However, we advocate such approaches, as qualitative inform-ation enriches quantitative studies and increases their communicability, while quantification tests the plausibility and consistency of qualitative studies. We also argue that normative elements in future studies are crucial when searching for futures that fulfil specific targets, and identifying viable strategies to reach these targets.

5 Contemporary future forest research

5.1 Introduction

Futures studies have a long tradition in the forest-based sector, and so-called forest sector outlook studies have been conducted regularly since the 1950s. The models used in these studies offer efficient means to unveil the past and present structures, particularly of markets for forest goods and products, although when used to generate long-term projections they generally neglect the possibility of structural changes (Hurmekoski and Hetemäki 2013). Another type of futures studies involves scenario analysis of the effects of forest management regimes. Such studies are increasingly common; a recent review identified 129 that examined management scenarios in Europe published during the last ten years (Hoogstra-Klein *et al.* 2016). The review showed that forest management scenarios are mostly simple, quantitative, and non-participatory in nature, but that there recently has been a shift towards longer spatial and temporal scales.

Regardless of the type of study involved, the basic categorization introduced in the previous chapter of probable, plausible, and preferable futures is valid. How these three approaches to the future may be used in future-oriented forest research is addressed in this chapter, by categorizing and synthesizing scientific articles emanating from the Swedish research programme Future Forests. This interdisciplinary and large-scale research programme illustrates recent attempts to explore forest futures with both traditional and more open, holistic approaches. The programme's current research is based on the principle that 'integrated knowledge deriving from a broad range of disciplines will be needed to seek effective solutions to complex forest governance and management problems that have no simple, or optimal solutions' (Future Forests 2013). However, the aim is not only to synthesize future studies in this programme, but also to use them to illustrate different types of forest-oriented futures studies and to discuss shortcomings and possibilities for refinements. However, the chapter starts with a brief historical overview of futures studies' development and connections to the Swedish forest arena's changing relationship with the future.

5.2 Past forest futures studies

In the early twentieth century a growing confidence in the state's capacity to govern the future, and the possibility of knowing the outcomes, in a distant future, of actions taken today, co-evolved with a battery of methods to enhance planning and foresight, such as inventories, surveys, and field trials (Mårald and Westholm 2016). It was important to counter effects of previous mistakes (e.g. agricultural misuse and industrial overexploitation), and in many ways the notion of a future was used for learning. Thus, during the forging of the first forest social contract, the question 'what did happen?' shaped the approach to the future, which at the time focused on putting things right, and thus 'what should happen'.

In the mid-twentieth century futures studies focused on progress and prediction; the future was seen as a product of governable technological progress (Kaijser and Tiberg 2000). Such perspectives also permeated Swedish forest-related forecasts and projections. For example, the governmental report that preceded a new Swedish Forest Conservation Act in 1948 stated that radical technological improvements were needed to meet future competition from other countries and from materials replacing wood (Betänkande 1946). Other investigations predicted a future shortfall in timber and were used to justify massive replacement of supposedly low-productivity and degraded stands with even-aged, high-productivity stands. These investments were made with the expectation that profits would be realized many decades into the future (Mårald and Westholm 2016).

The development of large-scale modelling generated discussion about the rapid growth in global demand for natural resources, and introduced a new element in the futures debate: the idea of limits to progress (Meadows *et al.* 1972; Warde and Sörlin 2015). This spurred new kinds of futures studies, a new era of strategic environmental research (Mårald and Westholm 2016), and establishment of several future-oriented institutes, such as the International Institute for Applied Systems Analysis (IIASA) in 1972 (Andersson 2012; Mårald and Westholm 2016). The Swedish governmental report *Forests for the future* predicted that forest industry consumption of forest materials would exceed domestic supplies (Skog för framtid 1978). However, the faith in progress governed by science and technology remained strong and underpinned an agenda to increase forest production to ensure supplies.

Post-war Swedish forestry, and accordingly the second social forest contract, illustrates how science and technology were used to reshape the present to fit a specific notion of the future and future needs (Mårald and Westholm 2016). 'What will happen?' was an important question in forward-looking forest analysis, which at the time focused on generating predictions regarding variables and conditions related to timber production and industrial development. However, Swedish forest politics were also shaped by a strong normative agenda largely driven by the question 'what should happen?'. For example, a primary intention of the Swedish Forest Conservation Act of 1979 was to ensure that appropriate silvicultural measures were taken in the present to ensure a steady supply of timber for the forest industry (a strongly desired future state). At the same time futures studies also adopted a more 'pessimistic' perspective. As outlined in the historical part of this book, from the 1960s onwards new forms of environmentalism increasingly challenged Swedish forestry and its objective to increase wood production. This involved a corresponding shift in which ways of knowing about the future were viewed as acceptable. Multiple futures were increasingly recognized, and dilemmas and political choice entered the futures debate (Att välja framtid 1972; Lönnroth *et al.* 1980). Faith in linear progress and the ability to steer society towards established goals was eroding; prosperity was increasingly linked to worries that extended generations into the future (Mårald and Westholm 2016). These trends were reflected in the Swedish forest debate of the 1980s and early 1990s, including contemporary efforts to explore the future. For example, the first broad interdisciplinary study of Swedish forest futures, by the Swedish Institute for Future Studies in the mid-1980s, criticized the forest industry's dominance in creating visions of the future, which the industry, according to the report, used to steer forest use solely to favour present-day profits (Wirén 1985).

Also during this period, the UN report *Our Common Future* explicitly linked the idea of sustainable development to future generations (Brundtland and Khalid 1987). Accordingly, present generations were assigned a transtemporal responsibility to ensure that economic activities were limited sufficiently to safeguard a range of values for the future (Langhelle 2000; Baker 2007). Moreover, the idea that the future can be deliberately steered towards clear and undisputed goals was increasingly challenged (e.g. Beck 1992; Giddens 1994). It is gradually being replaced by a more modest ambition to govern in the presence of complexity, risk, and uncertainty, acknowledging not only one or several futures, but also the possibility of unexpected developments (Lakoff 2007). Hence, complexity, uncertainty, and a need for integrated knowledge production are stressed in contemporary resource management and sustainability literature (Beland Lindahl and Westholm 2012).

These ideas are slowly but surely being adopted by the forest sector and associated international foresight and research institutions. The FAO Outlook studies, for example, increasingly promote multiple futures and choice. According to the FAO, contemporary outlook studies are intended to 'support policy reviews and strategic planning' by depicting 'the range of choices available to forestry policy makers' and describing 'the alternative scenarios that might arise as a result of these choices' (FAO 2011, What is an Outlook Study?, www.fao.org/forestry/outlook/en/). However, although the FAO's Outlook Studies of the 2000s reflect more holistic views (of the European forest sector, for example), critics argue that such studies still rely primarily on quantitative analysis of forest products, production, and trade (Pelli 2008). Consequently, they do not adequately address trends related to forest-based ecosystem services, social change, and human needs (i.e. Hurmekoski and Hetemäki 2013).

5.3 Present forest futures studies

To address this perceived deficiency in forest future studies various types of scenario analysis have been introduced in contemporary forest research, paving the way for a more complex and multifaceted engagement with possible forest futures reflecting increased awareness of uncertainty and the possibility of unexpected developments and shifts. The previously mentioned Future Forests research programme was motivated by recognition that forests are important components of socio-ecological systems that are being increasingly affected by climate change, globalization, and urbanization. During the period 2009–2016, various constellations of researchers rooted in diverse disciplines (natural sciences, social sciences, and humanities) have contributed to efforts to meet the programme's overarching aim: to assess the future of Swedish forests. Much forest-related research (including futures studies) is also done outside the Future Forests programme in Sweden, due to the economic importance of the forest industry. Nevertheless, here we focus solely on outcomes from Future Forests, as the programme's research plan contributed to a relevant and defined common purpose for all the studies included in the following analysis. Moreover, all the studies were conducted in a common interdisciplinary framework, enabling holistic exploration of silvicultural, ecological, social, and economic issues, and included trans-disciplinary collaboration with a broad range of stakeholder groups who have diverse norms and values regarding the future use of forests.

Future Forests has during its eight-year operating period published around 350 scientific articles in peer-reviewed journals, and numerous scientific reports, and has popularized articles in various journals and magazines (www.futureforests.se). Although most of these articles provide relevant information about future forest governance and management, the following analysis and synthesis are restricted to peer-reviewed scientific articles with an *explicit future outlook*, i.e. articles focused on issues that will influence developments during a specified future time period. Thirty-one studies meet this selection criterion, considering futures at spatial scales from local to global, covering timespans between twenty and 300 years into the future (Table 5.2). These thirty-one studies were categorized according to the framework by Börjesson *et al.* (2006) presented in Table 1.1, i.e. as studies of probable, possible, and preferable futures, and types of these overarching categories. Studies of probable futures were classified as either *Forecast* or *What if?* type; studies of possible futures as either *External* or *Strategic* type; and studies of preferable futures as either *Preserving* or *Transforming* type. We also classified the studies according to whether the applied approach was strictly qualitative or included quantification. The analytical focus, main statement on future development, and discussed consequences or recommended actions were also extracted from each study and are summarized in Table 5.1.

Table 5.1 The thirty-one studies with an explicit future outlook performed within the Future Forests programme, categorized according to the framework by Börjesson *et al.* (2006)

Future study	Spatial and temporal boundaries	Focus of analyses	Statement on future development	Consequences/recommended actions
Probable futures (forecast)				
Future Forests trends: Can we build on demographically based forecasts?[1]	Global, 2010–2050	Demographic effects on forest product demand	Increased world population and per capita incomes will generate large increases in forest product demand world wide	Consequences: risk for short-term forest depletion, but in the long term increased afforestation and intensification of forestry
Probable futures (What – if)				
Hydrological response to changing climate conditions: spatial streamflow variability in the boreal landscape[2]	Boreal forest landscape in north-Sweden, 2062–2090	Climate change effects on streamflow in boreal forest	Variability in streamflow will increase	Consequence: increased risks for extreme events (high and low) in forest streamflow
Possible futures (external)				
Impact of the 2°C target on global woody biomass use[3]	Global, 100 years	Demand for woody biomass to meet a 2°C limit for global warming	Woody biomass production will need to increase to meet the target for climate change	Consequence: production of woody biomass in short rotation forests in tropical climates will increase
* **Future governance: multiple challenges, diverging responses**[4]	Global and Nordic, 2010–2050	Institutional trends in forest governance	National governments' role and power to act in forest governance may decline at crucial times	Consequence: good forest governance depends upon democratic legitimacy in sovereign states and requires transparency and participation by all interests

Table 5.1 continued

Future study	Spatial and temporal boundaries	Focus of analyses	Statement on future development	Consequences/recommended actions
* Foresight on future demand for forest-based products and services[5]	Europe, 2010–2050	Drivers of demand for forest products and services	Free trade and global markets generate increasing regional differences in economic development	Consequence: greater regional differences in demand for forest products and services
Econometric modelling and projections of wood products demand, supply and trade in Europe[6]	Europe, 2010–2030	European country-specific projections of consumption, production and trade of wood products	Increased incomes generate increased consumption of wood products	Consequence: increased consumption of wood products
* Possible futures, future possibilities[7]	Sweden, 2010–2050	Drivers and trends affecting the forestry sector	Bioenergy use and the strength of political institutions influence future forest governance and management	Consequence: strong institutions and bioenergy replacing fossil energy contribute to a future in which sensible trade-offs about the forest are made
Global bioenergy scenarios – future forest development, land-use implications, and trade-offs[8]	Global, 2010–2050	Global changes in bioenergy and land use	Increased demand for energy wood	Recommended action: trade-offs between different ecosystem services will be required

Bioenergy futures: a global outlook on the implications on land use for forest-based feedstock production[9]	Global, 2010–2050	Global changes in bioenergy and land use	Increased demand for energy wood will not lead to forest loss but currently natural forest will need to be allocated to management	Consequence: to sufficiently protect biodiversity in sub-tropical and tropical areas extensive areas of temperate and boreal forest need to be managed
Global trends and possible future land use[10]	Global, 2010–2050	Global changes in land use	Global land scarcity is under way	Recommended action: the boreal forest industry must change its business concept to new products with higher value added and produce more from less
European Forest Sector Outlook Study II[11]	Europe, 2010–2030	European forests and the forest product markets	Increased demand for forest products and especially energy wood	Recommended action: trade-offs between different ecosystem services will be required
* The future use of Nordic forests: a global perspective[12]	Global and Nordic, 2010–2050	Global trends affecting future Nordic forest use	Increased demands on forests will sharpen conflicts between different interests	Recommended action: transition to address trade-offs to develop broad agreement on sustainable forest management
How to cope with changing demand conditions – the Swedish forest sector as a case study: an analysis of major drivers of change in the use of wood resources[13]	Sweden, 2010–2030	The industry and energy sectors' demands for woody biomass	Increased European demand for wood from Sweden for both material and energy uses	Recommended action: trade-offs between different ecosystem services will be required

Table 5.1 continued

Future study	Spatial and temporal boundaries	Focus of analyses	Statement on future development	Consequences/recommended actions
* **Trends and possible future developments in global product markets – implications for the Swedish forest sector**[14]	Sweden, 2010–2030	Globalization vs. regionalization and the degree of implementation of environmental policies are the two major drivers of demand for woody biomass in Sweden	Global development is the key driver of demand for Swedish woody biomass	The market for Swedish sawn wood appears bright, but the outlook for pulp and paper is less clear
Impacts of global climate change mitigation scenarios on forests and harvesting in Sweden[15]	Sweden, 2010–2100	Compares global demands for Swedish wood with national supplies	Increased global demand for wood from Sweden will mobilize the full harvesting potential	Recommended action: trade-offs between different ecosystem services will be required
* **Food, paper, wood, or energy? Global trends and future Swedish forest use**[16]	Sweden, 2010–2040	Trends affecting the forestry sector	Changes in energy systems, climate and forest politics, governance and global land use drive forestry sector transition	Recommended action: more integrated analysis of the forest/climate/energy nexus is needed to explore alternative pathways for future development
Effects of climate change on biomass production and substitution in north-central Sweden[17]	Sweden, 100 years	Climate change effects on forest growth and substitution potential	Climate change results in increased forest growth and substitution potential	Recommended action: increased forest growth should be used for substitution to maximize enhancement of forests' carbon balance

Possible futures (strategic)				
* Future forests: perceptions and strategies of key actors[18]	Sweden, 2010–2040	Key actors' perceptions of, and strategies for, the future; possible divisions between actors with implications for future forest use	Major division between actors who perceive biomass supply as unlimited and those who stress scarcity; frames conflict which may fuel future forest conflicts	Consequence: ideas related to social change, sustainability and justice are likely to become more salient and resulting power relations will determine whose preferred future will be realized
* Actors' perceptions and strategies: forests and pathways to sustainability[19]	Sweden, 2010–2040	Divisions between actors in the debate about future forest use and pathways to sustainability	Interest politics favour pathways that promote the status quo rather than change	Recommended action: expand the range of alternatives for the future and adjust asymmetrical power relations
Effects of intensified forestry on the landscape-scale extinction risk of dead wood dependent species[20]	Sweden, 250 years	Landscape-scale effects of intensive forestry on biodiversity	Application of intensive forestry to 50% of all spruce stands would increase risks for species extinctions, but this can be compensated by greater conservation considerations or larger set asides	Recommendation: TRIAD forestry can increase both production and biodiversity in a forest landscape
Modelled impact of Norway spruce logging residue extraction on biodiversity in Sweden[21]	Sweden, c. 100 years	Effects on biodiversity of extracting spruce logging residues	Extracting spruce logging residues has no significant effect on biodiversity	Recommended action: biodiversity is not the main issue for decisions on spruce logging residue extraction

Table 5.1 continued

Future study	Spatial and temporal boundaries	Focus of analyses	Statement on future development	Consequences/recommended actions
Effects of bioenergy extraction on dead-wood dependent species[22]	Sweden, 200 years	Effects on abundance of dead-wood dependent species in forest subjected to bioenergy extraction	Bioenergy extraction decreases the abundance of dead-wood dependent species	Recommended action: no stump and slash removal to protect rare dead-wood dependent species
Impacts of different forest management scenarios on forestry and reindeer husbandry[23]	Sweden, 100 years	Effects on the availability of reindeer forage under different silvicultural practices	Continuous cover forestry is better than clear-cutting forestry for promoting the future availability of reindeer forage	Recommended action: continuous cover forestry needed in the present to enhance future production of reindeer forage
Relative contributions of set-asides and tree retention to the long-term availability of key forest biodiversity structures at the landscape scale[24]	Sweden, 200 years	Effects of different retention practices on the availability of key structures for forest biodiversity	Tree retention practices implemented from the 1990s onward increase biodiversity of forest landscapes	Recommended action: maintain present tree retention practices to enhance forests' future biodiversity
Comparison of carbon balances between continuous-cover and clear-cut forestry in Sweden[25]	Sweden, 300 years	Carbon balances of forests subjected to different forestry practices	Clear-cutting forestry and continuous cover forestry have similar effects on forests' carbon balances	Recommended action: increased forest growth, regardless of the silvicultural practices, is needed to further improve forests' carbon balance

Potential effects of intensive forestry on biomass production and total carbon balance in north-central Sweden[26]	Sweden, 100 years	Carbon balances of forests subjected to intensive forestry	Intensive forestry results in increased forest growth and substitution potential	Recommended action: increased forest growth should be used for substitution to maximize enhancement of forests' carbon balance
Potential roles of Swedish forestry in the context of climate change mitigation[27]	Sweden, 2005–2105	Role of forestry in the national carbon balance	Increased forest growth and substitution contribute to climate change mitigation	Recommended action: intensify forestry to enhance its contribution to climate change mitigation
The potential role of forest management in Swedish scenarios towards climate neutrality by mid-century[28]	Sweden, 2010–2100	Role of forestry in reaching climate neutrality by mid-century	Increased forest productivity increases outputs of forest products and biofuels, and enhances ecosystem carbon sequestration	Recommended action: let Swedish forestry help efforts to cut national CO_2 emissions to sub-zero, thereby increasing scope to meet global targets
Simulation of the effect of intensive forest management on forest production in Sweden[29]	Sweden, 100 years	Forest growth and yield under intensified forestry	Intensified forestry increases forest yield	Recommended action: intensified forestry needed in the present to increase forests' future yield
Preferable futures (preserving)				
* Forest futures by Swedish students – developing a mind-mapping method for data collection[30]	Sweden, 20 years	Desired goods and services from the forest	Students from both social and natural science disciplines value forests' economic, ecological and social values	Consequence: forests should be managed to provide multiple ecosystem services

Table 5.1 continued

Future study	Spatial and temporal boundaries	Focus of analyses	Statement on future development	Consequences/recommended actions
Preferable futures (transforming)				
* Understanding consistencies and gaps between desired forest futures: An analysis of visions from stakeholder groups in Sweden[31]	Sweden, 40 years	Similarities in, and differences between, stakeholders' desired forest futures	Conflicts on forest management due to actors adhering to intrinsic vs. instrumental values of forests and advocating top-down vs. bottom-up forest governance	Recommended action: transparency and participation by all interests in development of forest policy

Each study's spatial and temporal boundaries, focus of analysis, main statement on future development, and discussed consequences or recommended actions are summarized. Studies marked with an asterisk (*) applied a strictly qualitative approach, while the others applied approaches including various quantitative methods.

In addition to the features used for the categorization of future studies applied in Table 5.1, there may be important differences in the target recipients. Here, we classify the studies according to whether they mainly target policy-makers or practitioners within the forest sector. One of the main differences between these two categories is that contexts of studies addressing practitioners are limited to the forestry sector, while those primarily addressing policy-makers span more societal sectors (Figure 5.1). Clearly, most of the Future Forests studies addressed policy-makers, and in most of those studies social science approaches and perspectives were applied. Notably, these approaches included quantitative and qualitative methods but in separate studies, using either econometric modelling or various techniques to investigate stakeholders' values, strategies, and power relations. The natural science studies mainly addressed forest sector practitioners and concerned future development of the forest landscape (Figure 5.1).

Moreover, as shown in Figure 5.1, most studies only involved academic experts, and generally applied quantitative methodologies, while there was

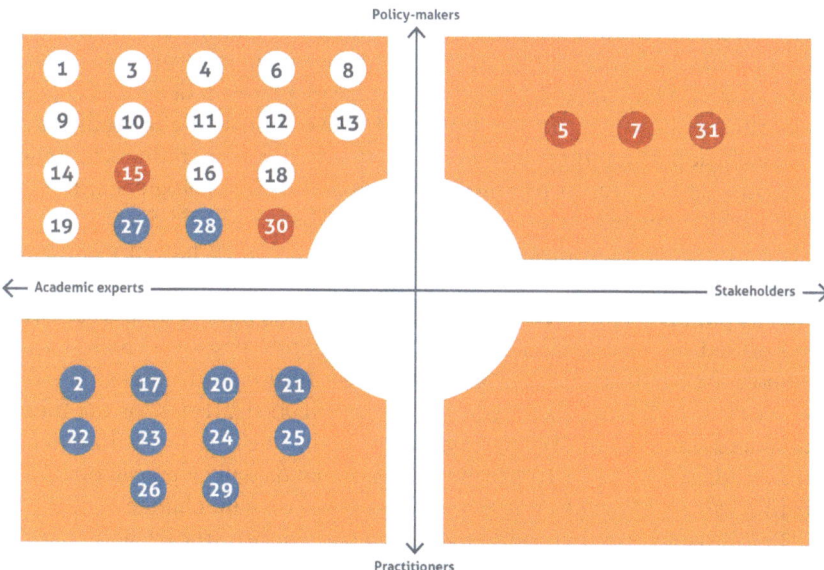

Figure 5.1 Positions of published futures studies conducted within the Future Forests programme in a two-dimensional framework showing whether they included only experts or also other kinds of stakeholders (with numbers of the latter indicated by distance along the axis) and whether they mainly addressed policy-makers or forest management practitioners (y axis).

The numbers refer to the studies summarized in Table 5.1 and listed in the notes on p. 84. Blue, grey, and orange numbers indicate natural science, social science, and interdisciplinary science studies (numbered as in Table 5.1 and the chapter endnotes), respectively. Graphical design: Jerker Lokrantz/Azote.

an inverse correlation between numbers of the other studies (which were more qualitative) and numbers of other kinds of stakeholders they included. Only a handful of the studies can be categorized as interdisciplinary in the sense that they attempted to combine frameworks based in natural and social sciences. Interestingly, all studies involving stakeholders can be classified as interdisciplinary and all were done on a level that provided better connections to policy-making than to practice (Figure 5.1).

The analysis shows that studies on possible futures have strongly dominated in the Future Forests programme. Out of thirty-one studies in total, twenty-seven were on possible futures and only four on either probable or preferable futures (Table 5.1). However, roughly equal numbers of studies of possible futures were classified as external and strategic types. The external studies focused on factors that were considered to influence the Swedish forest sector without the sector being able to influence them. They addressed the question 'How may changes in external factors affect the sector?', and in all these studies climate change was considered a major external driver of change. A basic common assumption underlying most of these studies was that society's efforts to mitigate climate change would result in increased demand for forest products and bioenergy on spatial scales ranging from global to national. End-point results of several of the studies also indicated that climate change will induce increases in demand for forest products and bioenergy in the future. However, most of the studies took this conclusion a step further by advising that better preparedness to handle trade-offs between forest production values and other ecological and social values will be needed, or even required in the future (Table 5.1).

The strategic future studies addressed the question 'What may happen if we act in a certain way?'. In contrast to the external future studies, which covered spatial scales ranging from global to national, all studies that could be classified as strategic covered a Swedish national scale (Table 5.1). A common aspect of all twelve of the strategic future studies was that they investigated or discussed consequences of specified strategies for the future. The shared context for all of these studies was the one laid out in the external future studies: that in the future increasing demands will be placed on forests due to society's efforts to mitigate and adapt to climate change.

There was a clear division between strategic future studies with a social science perspective, and those with a natural science perspective, so none of them could be classified as interdisciplinary across this divide. The studies with a social science perspective applied purely qualitative approaches and investigated development strategies for the future advocated by various stakeholders and the stakeholders' power relations, to assess their respective capacities to implement their advocated future strategies. Most of the studies addressed issues related to forest governance and identified major differences between strategies for the future laid out by different categories of stakeholders. A general finding was that stakeholders mainly concerned with forests' ecological and environmental values were more supportive of future strategies involving disruptive and transformative change of the current system than stakeholders who were mainly concerned with forests' production values (who favoured developments that could be largely accommodated in business-as-usual strategies). The studies also concluded that the current power relations between the stakeholder groups mainly favour the stakeholders promoting business-as-usual strategies. This indicates risks of the future being 'colonized' by these stakeholders, and closure of innovative and alternative pathways for future development suggested by less powerful stakeholders.

Natural science studies strongly outnumbered social science studies in the strategic category (accounting for ten out of twelve studies), and were mainly concerned with the future development of forest landscapes. They all applied a quantitative approach and mainly investigated effects of various forest management and conservation strategies on forest landscapes, but also considered their effects on Swedish society's capacity for climate neutrality, i.e. reduce net CO_2 emissions to zero (Table 5.1). Collectively, the twelve studies presented a multitude of strategies for the future, and quantified future development of forest landscapes with the aid of simulation techniques. The studies shared the same point of departure: that increasing demand for forest biomass would require intensification of forest management to enhance forest growth, which in turn would affect the forest landscape in various ways. Half of the studies addressed effects on forest landscape biodiversity or ecosystem services other than biomass production (e.g. production of lichens for reindeer forage) and half of them investigated the potential of various strategies of intensified forest management to increase future forest yields. A result highlighted in most of these studies is that forests are unlikely to be able to satisfy all desires and needs at the same time and everywhere.

As already mentioned, there were few studies on either probable or preferable futures in the Future Forests programme. Of the two studies on probable futures, one forecast future demand for forest products based on anticipated global demographic changes. The other projected streamflow changes due to climate change in a forest landscape in northern Sweden (Table 5.1).

Of the two studies on preferable forest futures, we categorized one as preserving, as it explored anticipated demand for forest goods and services in the future without questioning the present framework for forest governance and management. The other can be categorized as transforming, since it applied backcasting methodology to unveil a broad range of stakeholders' desirable futures (Table 5.1). Both studies of preferable futures applied strictly qualitative approaches.

5.4 Future forest perspectives

Futures studies are intended to support decisions today that have implications for the future. They can be powerful means to increase our preparedness to face the unknown and uncertain, but the key to engaging their full potential is the unveiling of multiple futures. Only then can future studies contribute to opening up new thinking and innovative pathways for future development. One way of doing this is to include several approaches to future studies within the same framework, as in the Future Forests programme. A strength of the programme is that it has made contributions to all three of the main categories of futures studies (probable, possible, and preferable futures). The programme has also included studies applying traditional forest sector outlook approaches and others applying integrated global modelling approaches (which have rarely explicitly focused on issues related to forest-land management). In addition, some Future Forests studies have applied purely qualitative approaches, for example backcasting, which have hardly ever been used in forest research.

Generally, the programme's studies have refrained from predicting 'what will happen'. Only two (of thirty-one) future studies make such claims, basing their predictions on slowly changing processes that are relatively easy to predict. For example, Malmberg (2015) extends historically verified relationships among demographic parameters, income, and demand for resources into the future, and predicts a dramatic rise in demand for forest products. He argues that demographic changes are likely to drive large-scale transformation of natural forests into intensive forest plantations (Malmberg, 2015). However, although the study is presented as a 'what will happen' forecast, the findings are conditionalized, acknowledging that technological shifts may decouple, or change, the historical relationships among the considered variables. Thus, the cited author argues, the presented trend should be regarded as a contribution to the pool of forecasts that can collectively help us to map an uncertain future (Malmberg 2015). This reasoning highlights the ambiguity of the boundary between studies of probable and possible futures.

The questions addressed by Malmberg (2015), regarding likely or possible effects of global demographic changes on future demands for forest products and bioenergy, were also addressed (along with effects of climate change), both quantitatively and qualitatively, in the 'what could happen' studies on possible futures. Integrated global modelling systems serving as powerful tools for theory-based quantitative analysis of the operating environment, and ultimately designed for tracking (and predicting) shifts in variables related to global sustainability, were used in some of these studies. The results include portrayals of possible futures where global demands for forest products and energy exceed supplies (e.g. Kraxner *et al.* 2013, Nordström *et al.* 2016, Lauri *et al.* 2017), calling for greater preparedness in policy processes to prioritize and make trade-offs among forests' values. The studies applying the more traditional forest sector outlook models came to the same overarching conclusion, although on more limited spatial scales – Europe (United Nations 2011) and Sweden (Jonsson 2011; 2013) – and underlined the importance of implementing global and European policy to mitigate climate change as a driver of increased demand for forest products and energy.

The studies indicate that this demand placed on forests may grow in the next twenty years, even under scenarios with intermediate levels of economic development (e.g. United Nations 2011; Jonsson 2012; 2013; Nordström *et al.* 2016). Several of the studies also suggest that high demands for forest products and bioenergy may result in forest management strategies that impair biodiversity and other ecosystem services (Jonsson 2013; Kraxner *et al.* 2013; Kraxner and Nordström 2015; Nordström *et al.* 2016). Moreover, in both Swedish- and European-level scenarios in which biodiversity protection is strengthened, i.e. the protected forest area is increased, the potential wood supply decreases significantly (United Nations 2011, Kraxner and Nordström 2015, Nordström *et al.* 2016).

Several of the qualitative studies of possible futures targeted the same dilemma: the mitigation of climate change driving increases in demands on forests. Restricting climate change to currently acceptable limits requires formulation and implementation of policies that facilitate replacement of fossil carbon in societies' energy and materials flows with biogenic carbon from forests. Otherwise, future demand for woody biomass may still increase, as a consequence of population growth and improvements in living standards, but the increases in its use will not result in climate change mitigation. Clearly, prioritization and trade-offs among forests' values will be needed to meet future challenges (Moen *et al.* 2010; Beland-Lindahl and Westholm 2011; 2012; Pelli and den Herder 2013) and such trade-offs must also consider the underlying drivers of the demands placed on forests. In order to enable just mechanisms for trade-offs, the importance of strong political institutions has been highlighted by several studies (e.g. Moen *et al.* 2010; Eckerberg 2015). Moen *et al.* (2010) also identify the strength of political institutions as one of two main drivers in a scenario-based analysis of future forest use, the other being the relative dependence on renewable and fossil energy (Figure 5.2).

The four possible futures suggested by Moen *et al.* (2010) all have different implications for the Swedish forest landscape. Forests' multi-use potential appears to be realized more fully by the regulatory and governance frameworks in the two proposed futures where political institutions are strong than in the futures where political institutions are weak. However, the main challenges concerning forests that the political institutions must address depend on whether use of renewable or fossil energy dominates. The impact of rampant climate warming is expected to be most challenging if fossil energy dominates, and trade-offs related to society's multiple needs for forests if renewable energy dominates. A common suggested feature of the two futures proposed to result from weak political institutions is that the market for forest products and energy is the main determinant of forest use. If fossil energy dominates, the forest may still be used for multiple purposes (including, for example, recreation, berry-picking, and hunting),

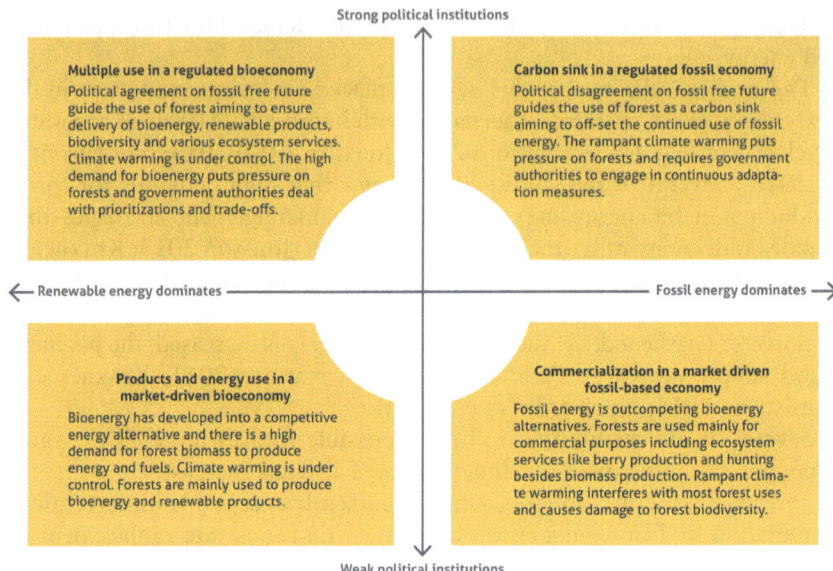

Figure 5.2 Four possible futures for the forest in 2050 as outlined in a scenario analysis involving both academic experts and stakeholders by Moen *et al.* (2010).

The four futures are framed by two main external drivers: the relative contributions of renewable and fossil energy to global energy consumption, and the strength (or weakness) of political institutions. Graphical design: Jerker Lokrantz/Azote.

but all these activities may become market-priced, and rampant climate warming may affect (in various ways) how they can be carried out. If renewable energy dominates, forests are expected to be used almost exclusively used for biomass production.

The correlation between future control of climate change and high demand for forest biomass for bioenergy indicated by Moen *et al.* (2010), as illustrated in Figure 2.2, is also predicted by other futures studies addressing external driving forces in the framework of the Future Forests programme (e.g. Nordström *et al.* 2016; Lauri *et al.* 2017). This putative connection has also been the point of departure for strategic futures studies of future forest management. These studies have mainly applied models that can be categorized as forestry decision support systems to explore possible forest futures under the influence of different management strategies. Several of the studies point out that some forest management strategies can contribute to mitigation of climate change by promoting carbon sequestration and the substitution of fossil materials and energy with forest raw materials and bioenergy, and that increasing growth and yields are crucial for realising this potential (e.g. Nilsson *et al.* 2011; Poudel *et al.* 2011; 2012; Lundmark *et al.* 2014; 2016; Cintas *et al.* 2016). Studies targeting strategies to sustain or enhance forest biodiversity clearly indicate that forest management practices designed to increase forest yields impair the biodiversity of forest landscapes and delivery of other ecosystem services (Dahlberg *et al.* 2011; Korosuo *et al.* 2014; Roberge *et al.* 2015; Johansson *et al.* 2016). Ranius and Roberge (2011) suggest strategies that can counteract such negative effects, but they are mainly operational at landscape scale, and thus generally challenging to apply in forest landscapes with highly fragmented ownership structures, like those in Sweden.

Of the thirty-one futures studies undertaken in the Future Forests programme, only two investigated preferable futures, using very different approaches. One (Nummelin *et al.* 2017) applied a new method for online construction of mind-maps envisioning preferred features of future forests, and the other (Sandström *et al.* 2016) a backcasting technique to unveil forest futures preferred by various forest sector stakeholder groups. The latter study confirms the well-known divide between Swedish forest sector stakeholders who prioritize ecological values and biodiversity protection versus those who prioritize wood and bioenergy feedstock production. Drawings produced by these two groups illustrating preferred futures for the Swedish forest in 2050 clearly reveal this major divide (Figure 5.3). Stakeholders promoting ecological values suggest that 50 per cent of Swedish forestland should be set aside for biodiversity purposes and not used for forestry wood production. In stark contrast, those who favour production suggest that Swedish forest growth should be increased by at least 50 per cent to enhance climate change mitigation. To reach this production potential several forest management measures were suggested, including large-scale fertilization of forests and drastic culling of the moose population (which currently causes severe browsing damage to forests) in rural areas, while increasing it in areas close to the large cities (Figure 5.3).

Although there were fundamental differences in the preferred future forests of the two stakeholder groups, there were also some similarities. For example, both thought that the number of forest-based companies should increase, and the biodiversity-promoting group even suggested that at least one million people should be employed in forest-based businesses in 2050; a massive increase from the current number (c 150,000). In addition, both groups regarded promotion of forest science and education, and the development of new high-technological products from forests, as important for the future (Figure 5.3).

Sandström *et al.* (2016) also identified futures desired by stakeholders that favour 'recreation and rural development'. These stakeholders are less well organized and influential than the two groups already mentioned. Indeed, the backcasting exercise revealed that an initial 'recreation and rural development' scenario they produced split, and two quite different scenarios evolved (Figure 5.4). One depicts a future in which rural development is based on traditional extractive forestry and wood-based product development. The other depicts a future in which development is based on forest

Figure 5.3 Drawings produced to visualize the contrasting forest futures in 2050 preferred by Swedish stakeholders favouring (A) forest biomass production and bioenergy, and (B) forest conservation and biodiversity protection (in both cases, representatives from around ten interest organizations).

Artist, Fredrik Saarkoppel.

Figure 5.4 Drawings produced to visualize the forest futures in 2050 preferred
by Swedish stakeholders (representatives of around ten interest
organizations) promoting recreation and rural development.

The stakeholder group envisioned a multi-use forest landscape, but did not agree about the
optimal governance mode; some advocated a 'top-down' governance mode with governmental
authorities having the ultimate power to make decisions on forest-related issues (A), while others
advocated a 'bottom-up' governance mode with forest owners having most power to make such
decisions (B).

Artist, Fredrik Saarkoppel.

Figure 5.5 Drawing produced to visualize the forest future in 2050 preferred by
Swedish Sami reindeer herding organizations.

Artist, Fredrik Saarkoppel.

aesthetics, tourism, and recreational values, so social values provide possible
future employment opportunities in rural areas. Both of these futures feature
landscapes with multiple-use forests, but the associated governance modes
fundamentally differed. The group promoting traditional extractive forestry
envisioned a 'bottom-up' governance mode with most of the ultimate power
to make forest-related decisions in the hands of forest owners. In contrast,
the group promoting mainly tourism and recreational values favoured a
'top-down' governance mode with governmental authorities having the
ultimate power to make such decisions (Figure 5.4).

Another group whose views were elicited in the study by Sandström
et al. (2016) is the indigenous Sami population, who have the right to herd
reindeers in Sweden. Their visions primarily focused on rights-based issues,
such as greater Sami property rights to land and water, implying crucial
power over decisions about forest land use and sales in the reindeer husbandry area, which covers approximately 50 per cent of Sweden (Sandström
et al. 2016). This would include (*inter alia*) protection of pastures used
for reindeer herding. In the reindeer husbandry area, reindeer herding would
have precedence over forestry and permitted forest management measures
would preserve or restore pasture land. In the envisioned scenario, old
trees would become more common in the forest landscape; there would be

no clear-felling; infrastructural elements such as roads and wind-turbine parks would be scarce and adapted to the needs of reindeer; and numbers of large carnivores would be significantly reduced. Consequently, Sami reindeer herding traditions would flourish and reindeer husbandry would be able to sustain reindeer herding communities throughout the reindeer herding area in Sweden (Figure 5.5).

5.5 Concluding remarks

There is increasing scientific interest in exploring the future of forests. In a recent review, Hoogstra-Klein *et al.* (2016) identified 129 scientific studies on future forest management in Europe published just in the preceding ten years. Similarly, Toppinen and Kuuluvainen (2010) identified and reviewed forty-seven modelling studies of the future development of the European forest sector and forest product markets published in a ten-year period. The thirty-one future studies done within the Future Forests programme provide illuminating examples of contemporary forest future studies framed by these two reviews. They include modelling studies at scales ranging from local and regional (with forestry decision support systems) through sectoral (with traditional outlook models) and global (with integrated modelling systems). They also include studies in which various qualitative approaches have been applied in scenario analysis and in investigations of stakeholders' normative visions of the future.

Whereas previous future studies often advocated just one preferable future at a time, we show that contemporary future studies offer multiple views of the future. Moreover, they apply diverse methods and adopt diverse perspectives, and so collectively they outline numerous alternative futures and broaden discussions about forest management and governance options to promote sustainable development. However, the analysis of the Future Forests studies also reveals shortcomings and gaps. For example, most of them were conducted separately, within established scientific traditions, and focused on distinct aspects of the future. There were few interdisciplinary efforts to bridge these divides, and none of the studies combined an ambitious qualitative approach with rigorous quantification. Moreover, few of the studies involved active interaction with stakeholders, although many of the social science studies investigated stakeholders' views and power relations. We also note that studies by social scientists appear to generally target policy-makers, possibly thereby avoiding engagement in practical forestry problems, whereas natural scientists mostly seem to target practitioners. In addition, none of the future studies were conducted in cooperation with stakeholders and focused on forest practice. However, such studies may be useful (or even essential) to enable local engagement and development of practical solutions to the global challenges we face (e.g. Milestad *et al.* 2014). Thus, when designing studies of future forest land use, there is clearly scope to improve the integration of established methods and involvement of stakeholders. This would be extremely valuable for further broadening of the spectra of alternative futures and associated management options.

Notes

1 Malmberg, B. (2015). In Westholm, E. *et al.* (eds). *The Future Use of Nordic Forests.* Heidelberg, New York, Dordrecht, London: Springer International Publishing.
2 Teutschebein, G. *et al.* (2015). *Water Resources Research*, 51, 9425–9446.
3 Lauri, P. *et al.* Submitted manuscript.
4 Eckerberg, K. (2015). In Westholm, E. *et al.* (eds). *The Future Use of Nordic Forests.* Heidelberg, New York, Dordrecht, London: Springer International Publishing.
5 Pelli, P., den Herder, M.(2013). EFI Report 87.
6 Jonsson, R. (2012). Geneva Timber and Forest Discussion Papers 59. UNECE/FAO.
7 Moen, J. *et al.* (2010). Future Forests (Report in Swedish).
8 Kraxner, F. *et al.* (2013). *Biomass and Bioenergy*, 57, 86–96.
9 Kraxner, F., Nordström, E. M. (2015). In Westholm, E. *et al.* (eds). *The Future Use of Nordic Forests.* Heidelberg, New York, Dordrecht, London: Springer International Publishing.
10 Nilsson, S. (2015).In Westholm, E. *et al.* (eds). *The Future Use of Nordic Forests.* Heidelberg, New York, Dordrecht, London: Springer International Publishing.
11 UN. (2011). European Forest Sector Outlook Study II. UNECE/FAO.
12 Westholm, E. *et al.* (eds). (2015). The Future Use of Nordic Forests. Heidelberg, New York, Dordrecht, London: Springer International Publishing.
13 Jonsson, R. (2013). *Canadian Journal of Forest Research*, 43, 405–418.
14 Jonsson, R. (2011). *Forests*, 2, 147–167.
15 Nordström, E. M., *et al.* (2016). *Canadian Journal of Forest Research*, 46(12), p 1427–1438. DOI: 10.1139/cjfr-2016-0122.
16 Beland-Lindahl, K., Westholm, E. (2011). *Forests*, 2, 51–65.
17 Poudel, BC. *et al.* (2011). *Biomass and Bioenergy*, 35, 4340–4355.
18 Beland-Lindahl, K., Westholm, E. (2012). *Scandinavian Journal of Forest Research*, 27(2), 154–163.
19 Beland-Lindahl, K. (2015). In Westholm, E. *et al.* (eds). *The Future Use of Nordic Forests.* Heidelberg, New York, Dordrecht, London: Springer International Publishing.
20 Ranius, T., Roberge, JM. (2011). *Biodiversity and Conservation*, 20, 2867–2882.
21 Dahlberg, A. *et al.* (2011). *Canadian Journal of Forest Research*, 41, 1220–1232.
22 Johansson, T. *et al.* (2016). *Forest Ecology and Management*, 371, 103–113.
23 Korosuo, A. *et al.* (2014). *Scandinavian Journal of Forest Research*, 29, 235–251.
24 Roberge, JM. *et al.* (2015). *Journal of Environmental Management*, 154, 284–292.
25 Lundmark, T. *et al.* (2016). *Ambio*, 45, 203–213.
26 Poudel, BC. *et al.* (2012). *Environmental Science and Policy*, 15, 106–124.
27 Lundmark, T. *et al.* (2014). *Forests*, 5, 557–578.
28 Cintas, O. *et al.* (2016). *Forest Ecology and Management*, 383, 73–84.
29 Nilsson, U. *et al.* (2011). *Forests* 2, 373–393.
30 Nummelin, T. *et al.* (2017). *Scandinavian Journal of Forest Research*, 32. Published online: 23 February 2017.
31 Sandström, C. *et al.* (2016). *Ambio*, 45,100–108.

6   Reflexive forest futures

6.1 Introduction

As outlined in the introduction of this book, transtemporal thinking recognizes that the temporal dimensions of the past and future are attached to the present, constantly affecting our understanding, choices, and trade-offs in real time. Thus, it raises awareness of the importance of responsible decision-making in the present, rather than postponing resolutions of problems. In this context, futures studies are vital for illuminating multiple options and perceptions of the future that may dissolve, or indicate potential ways to resolve, current complex problems. However, although they are important in enhancing our capacity in the present, they raise risks of colonising the future with predetermined solutions that reflect more the past than the future and may limit or diminish our ability to face the future. Hence, historical trends must be projected into the future cautiously.

For example, a feature of futures projected by many contemporary studies is the continuation of the longstanding conflict between forest production and nature conservation interests. Clearly, it can be argued that expected increases in demands for biomass and wood products may be met by an unquestioned consumption-and-growth paradigm that many studies project into the future. Moreover, envisioned threats to biodiversity and nature conservation are based on an undisputed paradigm of planetary limits that have already been breached and the expectation of future disaster (e.g. Rockström *et al.* 2009; Sardar 2010). Both arguments represent dominant ways of thinking that have shaped Swedish, and international, forest politics for decades. However, expectations that the dominance of these ways of thinking will continue (as indicated in many contemporary future studies) are based on an assumption that historical trends and existing relationships will continue to prevail, which is not necessarily true.

At the same time, it has been suggested that society and the forest sector face a transformative moment, a new kind of 'forest transition' (i.e. Mather 2001; Sandström and Sténs 2015; Westholm and Beland Lindahl 2015). Carbon, climate, energy, forests, and land use form a new kind of nexus around which alternative future trajectories may evolve. For example, a future 'bioeconomy' or 'green economy' is represented as a new option that would enable more efficient extraction of materials and energy from available resources, and thus synergistic solutions of several challenges (Beland Lindahl *et al.* 2015, 2016; Nilsson 2015). However, apparent novelty and change may conceal more or less deliberate strategies to maintain the status quo. Accordingly, there seems to be a major division between proponents of futures supporting 'equilibrium responses', involving at most moderate change, and those who call for more radical social change (e.g. Beland Lindahl and Westholm, 2011; Beland Lindahl *et al.* 2015). Moreover, stakeholders with a large potential to influence the future may prefer to reproduce the well-known highways of the past, because they are familiar or they will enable them to maintain their current activities and ways of life. To resolve associated problems, particularly the stark narrowing of options this entails, reflexivity is essential.

6.2 Reflexive forest future studies

An important insight of this part of the book is that *how* we investigate and envision the future will influence how the future influences our present decisions, so we should consider the future reflexively. This raises two obvious questions: what does reflexivity mean in this context, and what should characterize a reflexive approach to forest futures? These questions can be addressed by connecting to three of our central themes: transtemporal thinking, ways of knowing, and ways of doing.

First, transtemporal thinking, i.e. careful consideration of the temporal dimension, is essential in future studies. The further into the future the study extends, the more humble the research approach must be. Such transtemporal reflexivity highlights the open-endedness and inherent uncertainty of futures, and thus the great difficulty of projecting ongoing developments and present-day attitudes and values into the future. Nobody has access to any undisputable truth or 'road-map' for the future, so every perspective can contribute to envisioned futures. Moreover, careful consideration of whether to study *possible* futures or *preferable* futures, and thus how to include transtemporal reflexivity, is required in future studies. In studies of possible futures, the influence of the present is often more evident than in studies of preferable futures. The latter may even be done with an explicit intention to 'liberate' the future from the present. Transtemporal exploration of open futures should unveil more alternatives, thereby potentially generating more creative pathways and ways to dissolve, or resolve, current problems.

Second, is to acknowledge, and exploit, the multiplicity of ways of knowing futures. There are diverse cultures, knowledge systems, histories, aspirations for the future, and criteria for meeting those aspirations. By using a combination of complementary approaches and methods, a more multifaceted and richer picture of future challenges and options can be obtained. In addition, future studies can be improved by increasing their inter-disciplinarity. Indeed, interdisciplinarity is crucial to obtain *any* knowledge of the future (e.g. it is impossible to predict changes in forest growth rates without detailed predictions of climatic changes combined with know-ledge on how the forest may be managed and governed), and increasing the interdisciplinarity raises the potential to ability to cover the spectrum of alternative futures. In addition, transdisciplinarity must be encouraged, as it bridges the interfaces between scientific and other expertise in processes of problem-framing and problem-solving, thereby enriching outcomes of future studies. For example, combining quantitative modelling of biological material and physical systems with qualitative analyses of socio-economic and political systems is highly beneficial, although many methodological challenges remain to be solved. It is also valuable to combine normative and objective elements to promote transtemporal reflexivity.

Third, is to consider ethical aspects of the ways of doing future studies. The power to articulate and realize particular futures is unevenly distributed, and so are capacities to participate in future studies. Clearly, deliberating future alternatives is a critical component of a reflexive approach to forest futures, but only so far as the deliberating parties manage to imagine and take responsibility for the future consequences of their actions. Jasanoff (2003) suggests four questions that can be applied to most activities that aim to alter society: 'What is the purpose; who will be hurt; who benefits; and how can we know?'. Our knowing becomes reflexive when we grasp that our actions are not isolated, or isolatable, from their consequences, which may extend across substantial times and areas. Similarly, reflexivity in ways of doing future studies raises awareness of the need for responsible deliberation.

6.3 Concluding remarks

This part of the book discusses ways to understand and investigate the future, how future studies have been used in past and contemporary forest research, what we can know, what we can learn, and (finally) what constitutes a reflexive approach to multiple and uncertain forest futures. Many of the contemporary studies cited and discussed in this chapter suggest that changes in forest governance and management are needed to meet future challenges, in both Sweden and many forest-producing countries in the world. Thus, alternative pathways and forest futures must be carefully, and reflexively, considered.

Clearly, efforts to explore what these futures may entail are crucial for any rational attempt to make responsible decisions in the present. 'Looking forward' is an essential part of this process. There are tools to do this, but also a need for further methodological developments to improve and increase use of interdisciplinary, mixed-method, and participatory approaches. However, regardless of the approaches applied in future studies, non-reflexive explorations of the future may result in undesirable stereotyped patterns and structures to the detriment of future generations. Thus, there is a clear requirement for a reflexive approach to forest futures that can feed into the present: Reflexive Forestry, as explored in Part IV of the book.

Part III

Grasping the present

Revealing the present

The transtemporal thinking that permeates this book highlights the importance of choices and decisions made in the present, the transient time between historical developments and future projections. In Part I, we have shown that transtemporal thinking in different ways with varying emphasis on the past, present, and future, has infused the forest arena with different ideas and formed the basis for three broad forest social contracts. In Part II, a synthesis of future studies on forests and forestry shows that these types of studies, properly used, can play an increasingly important role in how we perceive and consider the future. Use of broader temporal perspectives, in studies with diverse spatial scales and contexts, clearly seems to enable identification of overall trends, and promotes both greater reflexivity and responsibility in relation to forest governance and management in situations with high ambivalence and uncertainty (Meadowcroft 2007). This seems especially important given that we seem to be about to enter a new fourth forest contract.

In order to govern change in situations characterized by ambivalence and uncertainty, a number of approaches to the governance and management of natural resources, including forests, have been developed. In theory and practice, these approaches are based on fundamental assumptions about how the managed systems function. An approach often includes normative ideas about human–nature relationships, which guide the design of analytical frameworks or schemes used to assess the success of ways of doing and progress towards specific goals. An approach also acts as a lens through which problems can be viewed and (thus) can strongly influence how pressing challenges are framed and handled (Rist and Moen 2013). Integrated approaches, including Sustainable Forest Management, claim to deal with challenges and trade-offs among ecosystem services and competing priorities of different actors when making decisions, and implementing forestry practices, by offering appropriate tools to meet stated objectives. In this manner, they also attempt to bridge an identified gap in the current forest social contract between governance and management. In the following chapters, we explore and compare a number of current mainstream approaches in terms of how they

(i)   encourage transtemporal thinking, reflecting past, presumable future, and current societal benefits and expectations related to forests;
(ii) integrate available ways of knowing to support the governance and management of forests;
(iii) define ways of doing forest governance and management, including the roles of actors/stakeholders.

Based on the findings of this comparative effort, we also explore how these approaches have been applied in a Swedish context, and which aspects may be important to bring about change in ways of doing, i.e. governance and management under a potential fourth forest contract. We start here by exploring the concept of Sustainable Forest Management and how it has been applied in a European context.

Ever since the Earth Summit in Rio de Janeiro, 1992, the concept of sustainable forest management (SFM) has been widely accepted as the main goal for forest policy and practice. Since then there has been a substantial proliferation (and convergence) of criteria and indicators to measure progress towards SFM through the Montreal process (MP), the Forest Europe process (FE, formerly Ministerial Conference on the Protection of Forests in Europe – MCPFE), and the International Tropical Timber Organisation (ITTO). At the global level there is, for example, a consensus in considering SFM as a balancing act between forest biodiversity conservation targets (MP C1; FE C4; ITTO C5; CBD Aichi Targets A3–4 and B7) and the maintenance of the productive capacity of forest ecosystems and their functions in watersheds and the global carbon cycle (Corezzola *et al.* 2016; Baycheva-Merger and Wolfslehner 2016).

In Europe, implementation of sustainable forest management builds on the endorsement of the Pan-European C&I for SFM by the forest ministers of the (forty-six) signatory countries of Forest Europe and the political foundation of the twenty-seven EU countries. The system is based on the seven thematic elements of SFM, as adopted by the UN, and consists of the following six criteria:

- maintenance and appropriate enhancement of forest resources and their contribution to global carbon cycles;
- maintenance of forest ecosystems' health and vitality;
- maintenance and promotion of forests' productive functions (wood and non-wood);
- maintenance, conservation, and appropriate enhancement of biological diversity in forest ecosystems;
- maintenance, conservation, and appropriate enhancement of protective functions in forest management (notably soil and water);
- maintenance of other socio-economic functions and conditions.

Sets of thirty-five quantitative and seventeen qualitative indicators have been developed from these criteria, which are intended to facilitate evaluation of progress towards goals associated with each criterion and the subsequent progress towards SFM overall. However, to date the pan-European set of criteria and indicators has primarily been used as a reference framework for dialogue and communication within the forest sector (although it has also been used for monitoring and reporting progress towards SFM, and to improve the comparability of forest information generated and disseminated by European countries). The concept has also been used as a reference framework for the development and adaptation of national policy instruments and/or forest-related policies (Linser *et al.* 2015). Despite the progress towards use of the concept and related indicators, various technical, financial, political, institutional, and sectoral barriers are hindering its implementation (MCPFE 2003; MCPFE 2007; Forest Europe 2011; Forest Europe 2015). With its primary focus on monitoring forests' various functions, the Sustainable Forest Management concept certainly plays an important role in highlighting their multifunctionality. Nevertheless, although it has been developed as a reference framework for dialogue and communication, it provides limited guidance for taking transtemporal thinking, ways of knowing, and ways of doing into account. Thus, there is a need to examine more closely other approaches and how they deal with these central aspects of a forest social contract.

7 Integrated approaches – in theory and practice

7.1 Introduction

Numerous complementary (and to some extent competing) approaches to the governance and management of natural resources have been developed, in parallel to Sustainable Forest Management, with the aim of providing guidance for the sustainable development of natural resources. Since there are so many, it would be impossible to cover them all here. Instead, we focus on the most prominent ones, which are claimed to take a system perspective and offer tools to guide decision-makers and practitioners, in different ways of doing, through the application of management methods or policy instruments. These are: the Ecosystem Approach, the Ecosystem-based Approach, Resilience Thinking, the Social-Ecological System Framework, the Ecosystem Service Approach, and Reflexive Governance. They are all also claimed to be holistic or integrative approaches to sustainability.

The Ecosystem Approach, which supports implementation of the Convention on Biodiversity (CBD 2004), developed more or less in parallel to the Ecosystem-based Approach (Waylen *et al.* 2014). Both have been applied in forest governance and management in Canada (Slocombe 1993; Price *et al.* 2009), in fisheries (Long *et al.* 2015), and climate change adaptation and mitigation studies (Naumann *et al.* 2011). In the last decade, Resilience Thinking has received rapidly increasing attention as a new paradigm in environmental and natural resource management. The approach is intended to illuminate how interacting systems of people and nature – or social-ecological systems – can best be understood and managed in the face of disturbances, surprises and uncertainty (Gunderson 2000; Folke 2006; Walker and Salt 2012).

The Socio-Ecological-System Framework (SESF) developed by Ostrom (2009; 2011) and colleagues (Ostrom 2007; Ostrom and Cox 2010; McGinnis and Ostrom 2014) is rooted in social science and the Institutional Analysis and Development (IAD) framework (Oakerson 1992). Like Resilience Thinking, the SESF focuses on the interaction between a resource system (including its constituent units) and the governance system and actors. The institutional heritage from the IAD-framework is mainly manifested in close attention to the governance system at multiple levels (Ostrom 2011). Another social science approach (rooted in economics) that has attracted interest in recent years as a framework for promoting the societal benefits of ecosystem conservation is the Ecosystem Service Approach. The approach has been strongly promoted by international initiatives such as the Millennium Ecosystem Assessment (MEA 2005) and The Economics of Ecosystems and Biodiversity (TEEB 2010), as well as by increasing application of ES-related policy instruments (e.g. Naidoo *et al* 2008).

Reflexive Governance, the final approach considered here, developed according to Voß and Bornemann (2011) from a combination of research traditions such as ecology (Holling 1973; Plummer and Armitage 2010), technology and innovation studies (Kemp 1994), and policy studies (Kenny and Meadowcroft 1999; Lafferty 2004; Voß *et al.* 2007). It has also evolved into a rather comprehensive approach, focusing particularly on transition management.

Despite the large number of theoretical approaches that have been presented (in addition to the six considered here), there are few empirically supported guidelines for applying them to enable transitions to a sustainable future.

Thus, in this chapter we compare similarities and differences among the six approaches, specifically focusing on the roles of transtemporal thinking, integrated knowledge, and the organization of governance to assess aspects that are consistent, complementary, or conflicting. To deepen understanding of the approaches we also compare their normative objectives. Then, with current challenges in mind, we ask how these approaches might assist or hinder governance and management in practice, but also to what extent they may promote a more reflexive approach to forestry.

7.2 The Ecosystem Approach

The phrase 'Ecosystem Approach' was first coined in the early 1980s, but was formally accepted at the Earth Summit in Rio in 1992, where it became a supporting concept of the Convention on Biological Diversity. It is defined as 'a strategy for the integrated management of land, water and living resources that promotes conservation and sustainable use in an equitable way' (CBD 1992). The strategy focuses on maintaining relationships and processes within ecosystems, and thus ecological integrity, setting targets for variables such as species composition and interactions among species (CBD 1992). As shown below, as one of the first comprehensive approaches to be developed, it has served as a role model or point of departure for the development of subsequent approaches.

The approach has been developed to include a toolbox based on a set of twelve principles, sometimes called the Malawi principles (see Box 7.1). Like the sustainable development concept that led to the SFM approach, the ecosystem approach has been disseminated through international networks and knowledge exchange. Although the CBD has been ratified by many countries, there is no uniform implementation of the Ecosystem Approach in the signatory countries. This can be largely attributed to variations in the nature and extent of previous policies and institutions, and 'the institutional configurations specific to each country' (Surel 2000).

The approach is often framed as being inclusive, holistic, and adaptive, with a bottom-up perspective (Box 7.1) that promotes various participatory and decentralized processes. This provides a way to put people at the centre of ecosystem management and thereby integrate learning and recognize diverse knowledge (Grumbine 1994). The approach also emphasizes the need for benefit sharing, thereby linking it to the more recently developed Ecosystem Service concept (MEA 2005). The Ecosystem Approach likewise recognizes uncertainty and the need for trade-offs between competing management objectives or between short- and long-term goals. Thus, it identifies adaptive management as an important element to achieve sustainable solutions (Walters and Holling 1990; CBD 2013). However, it does not fully acknowledge the need for trade-offs between production and conservation (an important normative issue), so application of the approach has been criticized for being both too ecocentric and too anthropocentric (Yaffe 1999). Hence, somewhat paradoxically, the approach has proved to be attractive to both the environmental movement, due to the holistic metaphors the concept resonates with, and to industrial capitalism due to its technical and mechanistic dimensions, which fit well with command and control style ideals. Thus, the 'contested and slippery character' of the approach makes it ambiguous and susceptible to discursive capture by competing narratives (Lucia 2015).

As the primary framework of action to be taken under the CBD, the overriding objective of the Ecosystem Approach is to protect biodiversity. However, the need to integrate it into agriculture, fisheries, forestry, and other production systems that affect biodiversity is acknowledged. In Sweden, the approach is primarily used for implementation of the CBD and protected area management (www.naturvardsverket.se/Stod-i-miljoarbetet/ Vagledningar/Ovriga-vagledningar/Ekosystemansatsen/). It has been applied with various degrees of success. For example, it has enabled some progress in the management of a few protected areas, which are generally subject to strongly top-down management (Holmberg *et al.* 2016; Hongslo *et al.* 2016). The approach has also been used in the establishment and management of model forests and biosphere reserves. However, in these attempted applications forests and forestry are considered primarily from an indirect perspective, so their management is not significantly affected (Schultz *et al.* 2012; Carlsson *et al.* 2015).

Box 7.1 The Malawi Principles

1   The objectives of management of land, water and living resources are a matter of societal choices.

2   Management should be decentralized to the lowest appropriate level.

3   Ecosystem managers should consider the effects (actual or potential) of their activities on adjacent and other ecosystems.

4   Recognizing potential gains from management, there is usually a need to understand and manage the ecosystem in an economic context.

5   Conservation of ecosystem structure and functioning, in order to maintain ecosystem services, should be a priority target of the ecosystem approach.

6   Ecosystems must be managed within the limits of their functioning.

7   The ecosystem approach should be undertaken at the appropriate spatial and temporal scales.

8   Recognizing the varying temporal scales and lag-effects that characterize ecosystem processes, objectives for ecosystem management should be set for the long term.

9   Management must recognize that change is inevitable.

10   The ecosystem approach should seek the appropriate balance between, and integration of, conservation and use of biological diversity.

11   The ecosystem approach should consider all forms of relevant information, including scientific and indigenous and local knowledge, innovations and practices.

12   The ecosystem approach should involve all relevant sectors of society and scientific disciplines.

CBD 1993

7.3 The Ecosystem-based Approach

As mentioned above, the Ecosystem-based Approach was developed in parallel to the Ecosystem Approach during the 1990s. Ecosystem-based management (EBM) is often presented as a comprehensive approach for improving the governance of the environment (Christensen *et al.* 1996; Slocombe 1993, 1994; Grumbine 1997; Yaffee 1999), which has been applied to both terrestrial and marine systems (e.g. CBD 1992; HELCOM 2008; Olsson *et al.* 2004). The approach was commonly defined as a paradigm shift within the domain of natural resource management, moving from a top-down, government-mandated, expert-driven approach to one that was bottom-up and based on collaborative governance and the integration of scientific and traditional knowledge (Meffe *et al.* 2002).

Thus, like the Ecosystem Approach the Ecosystem-based Approach (which can be seen as an attempt to codify basic elements of a new paradigm in resource management) is often considered to be holistic and to include collaboration with stakeholders. Moreover, governance modes such as public–private partnerships or adaptive co-management have been increasingly accepted as important components of the approach to achieve sustainable development (Glasbergen 2007, 2012; Van Huijstee *et al.* 2007) and adaptive, flexible implementation (Allen *et al.* 2011). Thus, the toolbox of the Ecosystem-based Approach (Box 7.2), with its focus on ecosystem integrity, participatory approaches, and adaptive management, overlaps substantially with the Malawi principles of the Ecosystem Approach (Box 7.1).

Although the 'principles' of the Ecosystem-based Approach are widely supported, there are wide variations in EBM definitions and applications. Consequently, the EBM is being implemented in numerous forms with various combinations of principles, possibly because (like the Ecosystem approach) it leaves room for various interpretations and orientations, ranging from anthropocentric to ecocentric (Yaffe 1999). This makes it difficult to instigate an EBM process and choose an approach that is most appropriate for a given area or environment. In combination with the lack of consensus regarding the key EBM principles, this creates a gap between theory and practice, which tends to impede successful application (Berkes 2011).

The approach forms the basis of the moose management system in Sweden. Forests provide important habitats and food resources for the

---

Box 7.2 Principles underpinning the Ecosystem-based Approach

1 Identification of the relevant ecosystems, and their boundaries and characteristics.
2 Agreement of management objectives for each ecosystem.
3 Development of long-term management objectives, as well as short- to medium-term objectives.
4 Establishment of sustainability indicators (including reference points, targets and limits) and the accompanying monitoring.
5 Application of a decentralised regional approach to management.
6 Better tailoring of research and information provision to support the ecosystem approach, including better knowledge of ecosystem interactions.
7 Application of Adaptive Management and the Precautionary Principle given the degree of uncertainty and dynamics of the ecosystem.
8 Effective enforcement.

Long *et al.* 2015; FAO 2003

moose population, but the moose can cause severe browsing damage. The rather hand-on tools of the approach have been used in efforts to develop greater knowledge-based and adaptive management of the moose, and to mitigate conflicts due to browsing damage (Sandström *et al.* 2013; Lindqvist *et al.* 2015).

7.4 Resilience Thinking

Ecological resilience was initially defined by Holling (1973) and the definition was subsequently refined, in more socio-ecological terms, by Walker *et al.* (2004) as 'the capacity of a system to absorb disturbance and reorganize while undergoing change so as to retain essentially the same function, structure, identity and feedbacks'. Thus, it is not a new concept, but 'Resilience Thinking' as a framework for achieving sustainability goals is a new and prominent extension (Folke *et al.* 2010; Walker *et al.* 2010). Interest in this approach has also been clearly seen in the context of forestry research and practice (Hughes *et al.* 2005; British Columbia Forestry Roundtable 2009; McAfee and de Camino 2010), including exploration of the challenges involved in applying the component ideas in a forestry or production system context (Rist *et al.* 2014)

Based on the view of social-ecological systems as complex adaptive systems, Resilience Thinking applies the concept of social-ecological resilience as a lens to address and understand dynamics within a system (Folke *et al.* 2016). A key objective is to identify factors that may either erode or enhance resilience. The acquired insights can then be applied when considering future options and strategies. The approach is also intended to foster complex adaptive thinking, learning, and broader participation through creating a shared understanding of the focal socio-ecological system and building trust (Walker and Salt 2012).

In terms of practical tools, Resilience Thinking embraces a collection of concepts that have been widely applied in individual case studies; for example, regime shifts, resilience thresholds, system transformation, adaptive cycles, and the social-ecological system perspective (all of which have been widely and usefully applied). In addition, the approach has been summarized in seven principles (Box 7.3), focusing on ways to build capacity to deal with unexpected change (Walker and Salt 2012, Biggs *et al.* 2015).

The last of the identified principles, polycentric governance, is a key feature of the approach. Polycentric governance is characterized by an organizational structure where multiple independent actors mutually order their relationships with one another under a general system of rules (Ostrom 1972). More specifically, it has been defined as a system consisting of

> (1) Multiple centers of decision-making authority with overlapping jurisdictions (2) which interact through a process of mutual adjustment during which they frequently establish new formal collaborations or

informal commitments, and (3) their interactions generate a regularized pattern of overarching social order which captures efficiencies of scale at all levels of aggregation, including providing a secure foundation for democratic self-governance.

(McGinnis 2016)

Recent advocates of the approach have increased consideration of what sustainability entails for human well-being, and added aspects of ecosystem-based stewardship as important elements of sustainable development (Folke *et al.* 2016). In summary, Resilience Thinking integrates a set of key concepts intended to provide an alternative way of thinking about social ecological systems and their management, but how this translates into policy or practice is still rather vague and undeveloped (Rist and Moen 2013).

Box 7.3 Principles underpinning Resilience Thinking

1 Maintain diversity and redundancy.
2 Manage connectivity.
3 Manage slow variables and feedbacks.
4 Foster complex adaptive systems thinking.
5 Encourage learning.
6 Broaden participation.
7 Promote polycentric governance systems.

Biggs *et al.* 2015

7.5 Social-Ecological System Framework

While the Ecosystem and Ecosystem-based Approaches, and Resilience Thinking, are rooted in natural sciences, the Social-Ecological System Framework has developed within social sciences. Consequently, in this framework sustainability is viewed as a societal challenge and (thus) solutions are also to be found in society. Based on the idea that all humanly used resources are embedded in complex, social-ecological systems (SESs), the SES framework was developed as a critical response to the myriads of simple theoretical models developed to analyze aspects of resource problems focusing on one-size-fits-all recommendations (Ostrom 2009; Ostrom and Cox 2010). The core idea of the framework is to diagnose why some SESs are sustainable while others collapse, through the identification and analysis of relationships at multiple levels of the complex systems across appropriate spatial and temporal scales (Ostrom 2009; 2011).

Instead of eliminating complexity, risk, and uncertainty, the idea is to learn how to dissect these aspects by focusing on the outcome of interactions between components of SESs, such as the resource system (e.g. a forest), resource units (e.g. trees), users (e.g. forest owners), and governance systems (e.g. organizations and rules that govern forests and forestry). The outcome is further assumed to affect the subsystems and their components, as well other larger or smaller SESs, through feedback loops (McGinnis and Ostrom 2014).

Use and development of this diagnostic approach are assumed to enable researchers to develop empirically supported answers to three broad questions (Ostrom 2007):

- What patterns of interactions and outcomes, such as overuse, conflict, collapse, stability, and increasing returns, are likely to result from using a particular set of rules for the governance, ownership, and use of a resource system and specific resource units in a specific technological, socioeconomic, and political environment?
- What is the likely endogenous development of different governance arrangements, use patterns, and outcomes with or without external financial inducements or imposed rules?
- How sensitive is a particular configuration of users, resource system, resource units, and governance system to external and internal disturbances?

A limitation of the SESF is that it does not outline a specific toolbox or set of sustainability assessment criteria (Partelow 2016). However, a set of seven design principles have been developed in relation to the IAD-framework, the forerunner to the SES approach, and are currently being developed to identify important characteristics of robust user- or self-organized systems. An eighth principle has been added to characterize the larger, more complex cases where, in a similar vein to Resilience Thinking, a polycentric governance perspective is promoted. In this case the idea of polycentricity is considered specifically appropriate to understand the governance of complex, modern societies, which requires institutional diversity embodied in multi-level, multi-purpose, multi-sectoral, and multi-functional units of governance.

However, a fundamental requirement for application of the concept is the capacity of groups to solve their own problems, i.e. to establish self-governance regimes (McGinnis 2016). This may or may not involve formal governance bodies. However, McGinnis and Ostrom (2011) find that, in policy-making negotiations, citizens and officials often collaborate in various forms of partnerships to develop or design solutions to identified challenges. The process requires experimentation and an entrepreneurial approach, so it is closely linked to the ideas of adaptive management.

The eight principles, which are highly generic, can be used to diagnose problems and prospects associated with current situations and focus to a large extent on institutional factors, such as the rules and regulations

Box 7.4 Eight principles underpinning the SES Framework

1   Clearly defined boundaries (clear definition of the contents of the system and effective exclusion of external un-entitled parties).
2   Rules regarding the appropriation and provision of common resources that are adapted to local conditions.
3   Collective-choice arrangements that allow most resource appropriators to participate in the decision-making process.
4   Effective monitoring by representatives of the appropriators.
5   A scale of graduated sanctions for resource appropriators who violate community rules.
6   Mechanisms of conflict resolution that are cheap and easy to access.
7   Self-determination of the community recognized by higher-level authorities.
8   In larger SESs, organization in the form of multiple layers of nested enterprises and polycentric governance, with small local SESs at the base level.

Ostrom 2009

governing social and ecological systems. The approach has been widely applied in studies that have explored the complex interactions between local communities and their forests in many countries (Binder *et al.* 2013; Sandström *et al* 2013).

7.6 Ecosystem services approach

The concept of ecosystem services (ES) has gained global attention in recent years as a framework for promoting the societal benefits of ecosystem conservation. As described in Part I, the idea that forests have wider 'functions', such as regulating the climate and watersheds, in addition to providing direct human benefits, dates back to the nineteenth century. However, the ecosystems services concept has made this idea more salient and widened the spectrum of benefits. This trend has been strongly influenced by widely read scientific publications and international initiatives, including the Millennium Ecosystem Assessment (MA 2005) and The Economics of Ecosystems and Biodiversity initiative (TEEB; Kumar 2010), as well as by increasing application of ES-related policy instruments (e.g. Naidoo *et al.* 2008). The Millennium Ecosystem Assessment (MEA 2005) defined ecosystem services simply as 'the benefits people obtain from ecosystems', both natural and managed. Several categories of services are commonly recognized: provisional, regulative, cultural, and supporting. The first three categories directly affect human well-being, while the latter has an indirect impact by supporting the first three (Costanza *et al.* 1997; Daily *et al.* 1997; Wall 2004; MEA 2003, 2005).

Policy directives of many governments around the world include integrated goals to protect ecosystem services. For example, the governments of China, Costa Rica, Mexico, and Ecuador have all established schemes to pay landholders who engage in management (e.g. protection of forest or improved agricultural practices) that secures the supply of hydrological services (e.g. Sánchez-Azofeifa *et al.* 2007; Liu *et al.* 2008). In Sweden, the government adopted a strategy for strengthening biodiversity and securing ecosystem services in 2014 ('A Swedish strategy for biodiversity and eco-system services', Govt. Prop. 2013/14:141). This contributes to long-term Swedish nature conservation policy connected to the Swedish environmental quality objectives, the generational goal, targets in the EU Biodiversity Strategy to 2020, and the international Aichi Biodiversity Targets set in the UN Convention on Biological Diversity (CBD).

However, attempts to integrate the approach into Swedish forestry management and governance faces challenges associated with the Swedish model and policy history (Beland Lindahl *et al.* 2015). A preference for ecological 'modernization' (perceiving possibilities of both maintaining economic growth and meeting environmental objectives through technical innovations, green design, and environmental reforms) has promoted a 'win-win' view that conflicts with the integrations and trade-offs associated with a more ecosystem services-oriented approach (e.g. Daw *et al* 2015). Exploration within a framework that names and categorizes non-production forest values has more clearly highlighted the reluctance to recognize limits or goal conflicts in Swedish policy, and a lack of mechanisms to enable trade-offs and choices.

The concept particularly highlights the key problem linked to trade-offs and interactions; management activities intended to enhance the provision of a particular set of prioritized ecosystem services may inadvertently impair other services that are vital for the system's operation. Such trade-offs may be related to straightforward distributional impacts of management, while others may arise from social-ecological complexity. However, use of the ecosystem services concept in discussions, with a sustainability or resilience perspective, has more clearly exposed systems' high dependence on sup-porting and regulating services, and costs of externalities imposed on other systems and locations (Rist and Moen 2013; Rist *et al.* 2014). Reconciling these insights with a strong economic focus and existing distributions of power among actors and institutions in the forest sector is a major current challenge.

To aid efforts to meet such challenges, to value environmental services more rigorously, to provide evidence to identify appropriate management options to optimize public benefits across the breadth of ecosystem services, and to avoid potentially substantial costs and risks arising from overlook-ing implications for some services, specific principles have been developed

(Box 7.5). These include not only economic principles, but also others related to ethics, participatory, and scientifically robust methods and Resilience Thinking.

The increasing attention paid to ecosystem services can be attributed largely to the concept's potential to broaden appreciation of the contribution of ecosystems to human well-being. However, while the ecosystem services concept supports interdisciplinary dialogue (Cowling *et al* 2008), and highlights specific society-ecosystem interactions, it has mainly been used to bridge the gap between ecological and economic considerations (Daily 1997; Millennium Ecosystem Assessment 2005) and thus to date primarily

---

Box 7.5 Principles underpinning Ecosystem Service Assessment

1 Articulate a clear purpose for the assessment and a rationale for the methods used.
2 Reflect a fair and honest effort to represent ecosystems and all of the benefits they provide without intent to produce a predetermined outcome.
3 Identify and engage all interested and affected stakeholders in a transparent, inclusive manner.
4 Use interdisciplinary approaches to address the landscape attributes, ecological functions, and stakeholder perspectives at scales that allow decision makers to understand the full range of benefits, costs, and potential solutions.
5 Assess the full suite of ecological, social, and economic costs and benefits in quantitative and qualitative terms using credible methods, while avoiding the double counting of monetized values.
6 Consider resilience and the ability to maintain biodiversity and sustain ecosystems for current and future generations.
7 Base assessment on the best scientific information available while disclosing uncertainties that bear on the decision, and provide analysis on the potential effects of those uncertainties.
8 Apply robust methodologies and approaches that strive to be consistent, repeatable, and transparent, while encouraging the improvement of ecosystem services assessment methodologies and tools.
9 Provide a rationale for excluding any social, ecological, or economic attributes relevant to the management decisions that were not included in the assessment, and make the full assessment available for technical review.
10 Use language that is relevant to the intended audience and sparing in its use of acronyms and abbreviations to make valuation results accessible for non-technical stakeholders.

incorporates these two perspectives. This includes a dominant focus on monetary valuation (Abson *et al.* 2014) and the increasingly popular 'payments-for-ecosystem-services' (PES) schemes (Engel *et al.* 2008; Wunder *et al.* 2008). Accordingly the concept has been criticized, for example for its anthropocentric focus or for contributing to commodification of nature (Schröter *et al.* 2014).

7.7 Reflexive Governance

With its roots in complex systems thinking, the Reflexive Governance approach is conceptualized as both a condition for governance and a specific strategic orientation resulting from this condition (Loorbach 2007; Voß and Bornemann 2011). Inspired by the concept of reflexive modernity (modernity, with reflection on the challenges and opportunities it raises), reflexive governance refers to governance that involves reflection on the challenges, opportunities, and responsibilities it engenders (Beck 1992).

Hence, in contrast to many other approaches, which assume governance to have an external position with the intention to shape society and technology in order to maintain desired socio-ecological systems, the Reflexive Governance approach recognizes that governance is often part of the problem. Thus, as expressed by Voß and Bornemann (2011): 'the reflexive stance toward governance abandons the assumption of "one" adequate problem framing, "one" true prognosis of consequences, and "one" best way to go that could be identified in an objective manner from a neutral, supervisory outlook on the (social–ecological) system as a whole.' Consequently, Reflexive Governance includes a learning component, involving continuous self-oriented examination of positive and negative outcomes (substantive dimensions, sometimes called single-loop learning), and the relation of governance itself to these outcomes (structural/procedural dimensions; sometimes called 'double-loop learning') (Boström *et al.* 2017).

This, in turn, has implications for the definition of sustainability. Here, it is not considered as an end-state, but understood as the capacity of society to learn about the conditions of its future existence and wants. The approach also incorporates a specific problem-solving tool – transition management – to integrate diverse perspectives, expectations, and strategies in a complex understanding of societal change (first-order and second-order reflexivity). It emphasizes the interlinkage of problems and scales, as well as long-term and indirect effects of various actions. Transition management has been particularly applied in the Netherlands in initiatives to shape innovation processes and structural transitions in complex socio-technical systems such as those directly involved in energy provision, mobility, and agriculture, but only indirectly in forests and forestry to date (Kemp *et al.* 2007a; VROM 2001).

Transition management builds on the idea of co-evolutionary socio-technical systems, where structural transitions evolve in a particular temporal pattern (Geels 2002). Transition management provides general principles, such as systems thinking, long-term thinking through historical examples, scenario analysis, an adaptive orientation towards learning, innovation, and experimentation, as well as participation (Kemp *et al.* 2007a; Rotmans *et al.* 2001). These principles are applied in an operational governance design that combines the establishment of a transition arena to facilitate interaction and learning in a proactive manner with the development of systemic transition instruments such as sustainability visions, mobilization of actors, initiation of projects and experiments, and continuous monitoring and evaluation of the processes (Kemp and Loorbach 2006; Kemp *et al.* 2007b).

To consider long-term frames (both normative and explorative), historical case studies and scenario development and analysis have come to play important roles in transition management (Rotmans 2005; Loorbach 2007; Geels 2005; Nordlund *et al.* 2011). The scenarios are often developed in three steps to take into consideration *'multi-level criteria'* or elements of change in combination with uncertainties that can be identified in the present, thereby recognizing and investigating potential drivers of long-term transitions. By adding an analysis of *multi-pattern criteria* or the interactions between actors and structures of the system, the patterns that may promote or hinder transition can be identified. Finally, the *multi-phase criteria* focus

---

Box 7.6 Principles underpinning Reflexive Governance

1 System-thinking in terms of more than one domain (multi-domain) and different actors (multi-actor) at different scale levels (multi-level).

2 Long-term thinking (at least twenty-five years) as a framework for shaping short-term policy.

3 Backcasting and forecasting: the setting of short-term and longer-term goals based on long-term sustainability visions, scenario studies, trend analysis, and short-term possibilities.

4 Strategies and strategic experiments to actively deal with uncertainty through transition management and promote learning about a variety of options.

5 Integrated production of knowledge about problems and their dynamics, including diverse scientific disciplines and practice perspectives, to get more robust knowledge and strategies based on shared understanding.

6 Interactive strategy development by actors with various sources of influence.

7 Iterative, participatory formulation of governance objectives, taking account of diverse and changing social values.

Voß *et al.* 2005; Loorbach 2006; Loorbach, 2010

on timings and the four stages involved in transition management: the predevelopment, take-off, acceleration, and stabilization phases. The nature and speeds of these phases inform the character of the transition process (Sondeijker 2009).

The Reflexive Governance approach has attracted extensive interest in research, but also criticism, particularly regarding its political and democratic implications (Rotmans and Kemp 2008). Although Boström *et al.* (2017) consider reflexivity to be a useful analytical concept, they are more cautious about its usefulness in practice. The authors claim that the concept needs to be 'used with caution, put in context, and with a firm systematic look at its boundaries and opposites'. Moreover, they highlight the need to address power relations to understand both actors promoting change as well as those who aim to prevent change.

7.8 Comparing integrated approaches – main similarities and differences

Although the six integrated approaches considered in this comparison had widely differing roots (in natural sciences, social sciences, or schools of science and technology studies), they have substantial similarities. This especially applies to their framing of the ways of knowing and focus on ways of doing through, for example, collaborative decision-making and iterative management or learning by doing. All of the approaches promote the integration of various knowledge systems. However, they all provide limited guidance on how to achieve such integration in practice. They all also advocate the inclusion of all actors with a stake in natural resource governance and management, and stress the importance of systems thinking, considering human–nature interactions, and acknowledging inherent uncertainty.

Clearly, developers of these approaches have been aware of (and inspired) by the other approaches, although they all have distinctive features or emphasize different aspects. Thus, the comparison indicates several significant differences in the six integrated approaches – the Ecosystem Approach (AM); Ecosystem-based Management (EBM); Resilience Thinking (RT); Socio-Ecological System Framework (SESF); and Reflexive Governance (RG) – as summarized in Table 7.1 and outlined below.

The first differences are in the approaches' articulation of normative objective and management targets. Four of the approaches focus on various forms of maintaining a certain state through fostering ecological integrity (EA/EBM), the resilience of systems to return to their original states after disturbance (RT), or robustness, i.e. the ability of a system to reach a desired future state despite uncertain future events or unexpected developments (SESF).

Generally, the focus is less explicitly stated in the Ecosystem Service Approach than in the other approaches. However, in the Millennium

Table 7.1 Comparison of integrated approaches

	EBM/EM	Resilience thinking	SESF	ESS	Reflexive governance
Scientific origin	Natural science	Natural science	Social science	Social science	Socio-technology
Normative objective	Ecological integrity	Resilience	Robustness	Sustainable use of biodiversity and ecosystem services	Sustainable development as a process of social change
Transtemporal thinking	Implicitly future-oriented	Explicitly future-oriented	Implicitly future-oriented	Implicitly future-oriented	Explicitly future-oriented
Ways of knowing	Integrated knowledge production	Integrated knowledge production	Integrated knowledge production	Integrated knowledge production	Integrated knowledge production
Ways of doing	Decentralized	Polycentric	Polycentric	–	Reflexive
Actors	All relevant sectors of society and scientific disciplines	A variety of participants	Resource appropriators	All interested and affected stakeholders	Relevant actors
Management	Adaptive management	Complex adaptive systems, or self-organising systems	Self-organising systems	Adaptive management	Transition management
Means to gain directionality/toolbox	Principles and operational guidance	Resilience assessment	Diagnostic assessment	Ecosystem service assessment	Interactive strategy/portfolio management

Ecosystem Assessment the target is expressed in terms of the sustainable use of biodiversity and ecosystem services for the current and future benefit of people (MEA 2005). The Intergovernmental Panel on Biodiversity and Ecosystem Services (IPBES) uses the concept 'Nature's contribution to people' when referring to all ecosystem goods and services, separately or in bundles, that contribute to the benefits that humanity obtains from nature. Here it is also acknowledged that all nature's benefits have anthropocentric value, including instrumental values – the direct and indirect contributions of ecosystem services to a good quality of life (Decision IPBES-2/4).

In contrast to the other approaches, Reflexive Governance, with its focus on sustainable development as a process of social change, has a fundamentally different objective from the other approaches. Rather than concentrating on maintaining the considered system, the aim of Reflexive Governance is to help decision-makers to influence and organize evolutionary processes of societal change (Kemp and Loorbach 2006).

The second set of differences lie in the approaches' transtemporality. None of them promote pure transtemporal thinking, i.e. seeing the past, present, and future as interlinked. With the focus on sustainable development, all six approaches are essentially future-oriented. However, two of the approaches include temporal aspects and tools to address issues spanning multiple timescales. Resilience Thinking encourages structured processes such as scenario planning to explore and evaluate alternatives to account for change and uncertainty. Closest to transtemporal thinking is the Reflexive Governance approach, which acknowledges the conflict between long-term imperatives and short-term concerns (Kemp and Loorbach 2006), one of the most crucial challenges that policy-makers who deal with governance face on a daily basis. Thus, the approach includes methods such as backcasting and forecasting to enable short-term and long-term goals to be set, based on long-term sustainability visions, scenario studies, trend analysis, and short-term possibilities, as discussed in Part II.

The third set of differences is in the ways of doing, i.e governance modes and tools or management procedures, the approaches promoted to achieve policy implementation. While the Ecosystem Approach and the Ecosystem-based Approach promote decentralized management, Resilience Thinking and the SESF advocate a combination of polycentric governance with adaptive co-management or local and adaptive co-governance arrangements, respectively. As indicated by its name, Reflexive Governance promotes a reflexive governance mode to meet contemporary and future challenges where policy processes are as much about shared problem construction as they are about collective solutions. The differences in governance modes advocated by the approaches probably reflect their development and adaptation to changes in existing modes of governance as much as normative governance preferences.

Regarding management, all the approaches promote the application of Adaptive Management (AM) in some way, except Reflexive Governance, which advocates Transition Management (TM). AM and TM have many common components, such as recognition of the need to design management processes with sufficient flexibility to respond to the uncertainties and dynamics of complex systems. By addressing these challenges, both management modes promote an experimental orientation, a learning-by-doing approach, as well as the participation and collaboration of diverse actors in order to integrate multiple perspectives and resources for governing (for a comparative overview of the two, see Voß *et al* 2011). However, while AM is used as a method to maintain (for example) robustness and resilience, TM builds on the idea that, with appropriate management, the transition dynamics of societal systems may lead to establishment of a higher order of organization, complexity, and (hence) new governance or management regimes (Rotmans 2005).

Thus, AM involves adjustment to changes in the structure of a system, while TM involves directing and guiding in anticipation, often based on scenario analysis, of future changes in the system. There are also differences in the management process. In AM (and adaptive governance) a cyclical plan is often developed, including a combination of short-term and long-term steps to tackle uncertainty (Rist *et al*. 2015). According to Loorbach and Rotmans (2006) this is a 'low risk strategy' that can easily lead to 'no regret' strategies, i.e. strategies that will do little damage. In contrast, TM often encompasses a portfolio of experiments as a basis for testing and learning.

In all the approaches, a set of principles or guidelines has been developed to gain directionality for generating common decision-making processes and action, but some are more proactive, prescriptive, and diagnostic than others when suggesting methods or tools to achieve overarching objectives. In the EA/EBM approaches, stated principles and operational guidance are used proactively to enhance benefit sharing, AM, decentralization, and local management action, together with a focus on intersectoral cooperation. Resilience assessment is promoted as a key method in Resilience Thinking (Walker and Salt 2012; Biggs *et al* 2015). This approach also offers proactive and rather normative advice for promoting progress through its focus on ways to maintain diversity, manage connectivity, and foster complex adaptive systems thinking, learning, and participation. Reflexive Governance is the approach with the most strongly proactive perspective, but instead of promoting a single solution (ecosystem integrity or resilience) it advocates portfolio management, i.e. offering a mix of policy alternatives, instrument and tools.

As mentioned above, the SESF is based on a diagnostic and rather reactive approach. Thus, it is more of an analytical tool for identifying strengths and weaknesses of focal systems than a normative approach for attaining a desired end-state. Similarly, the Ecosystem Service Approach proposes analytical tools and methods for valuing ecosystem services. In principle, these two approaches could be used to complement any of the other approaches.

7.9 Concluding remarks

The integrated approaches compared here can be viewed as points along a path of increasing development in ideas about optimal ways to tackle the complex tasks of natural resource management and bridging the gap between governance and management (Rist and Moen 2013). However, the ideals of one approach are rarely achieved in practice and, as the inherent problems become evident over time, challenges by other ideals and paradigms induce refinements. There are also other sources of change, including new knowledge, changes in public values, and political dynamics. Thus, the approaches have not developed in an entirely linear fashion, along a clear continuum (Rist and Moen 2013). In the following section, we analyze the extents to which these approaches have been applied in a Swedish context and have contributed to development of the Swedish forestry model.

8 Efforts to bridge governance and management in Swedish forests

8.1 Introduction

Although Swedish forestry has been discussed and analyzed on the basis of almost all the approaches listed and described above (Rist and Moen 2013), none of them has been directly applied to forestry. Only parts of the approaches have influenced parts of Swedish forest policy and (thus) the ways of doing through governance and management of forests. Instead, during the last two decades forest policy and governance have developed incrementally within the framework of the Swedish or Nordic Forestry Model, what we in this book have defined as the third social contract. This bottom-up approach places responsibility for meeting and trading-off the multiple goals related to forest services on the individual forest owner or forest manager (Gustafsson *et al.* 2012). Moreover, the model has been labelled 'more-of-everything' (i.e. Beland Lindahl *et al.* 2015), due to the optimistic view that existing resources can meet all of the goals simultaneously. However, as initially mentioned, an explosion of policy goals has reduced this possibility, and the model heavily relies on voluntary implementation instruments, with weak mechanisms to make trade-offs. This weakness has been defined as an implementation deficit, which tends to undermine the legitimacy of the third social contract.

Hence, instead of integrating governance and management, the model appears rather disjointed and misconceived, due to the lack of links between policy objectives and management practices, which makes it vulnerable to criticism from both the industrial forestry sector and the environmental movement. Thus, like the lack of forest policy integration at the European level, where there are strikingly diverse understandings of forests and their management (Winkel and Sotirov 2014), this exposes the model to indirect influences from other sectors and approaches. For example, Swedish nature conservation and moose management policies – respectively based on the Ecosystem and Ecosystem-based Approaches, both of which promote ecological integrity (see Boxes 7.1 and 7.2) – indirectly influence forest policy through policy objectives and management structures (Sandström *et al.* 2013). Furthermore, the Swedish Government has recently decided to adopt a strategy for strengthening biodiversity and securing ecosystem services (Bill 2013) based on the Ecosystem Services Approach (see Box 7.5).

Consequently, with the lack of an integrated approach for making efficient and legitimate adjustments, trade-offs, and choices in the process of implementing policy into management practice, there will inevitably be goal conflicts. This can be illustrated, for instance, by the conflict between the Production and Environmental objectives in the Swedish forest policy (Beland Lindahl 2008), between the Production objective and demands for considering Sami reindeer husbandry (Widmark 2009), and requirements to protect cultural heritage sites (Sandström and Lindkvist 2009). In response, efforts have been focused on developing forest management practices to handle these conflicts and manage trade-offs without properly addressing the perceived disconnect from the governance system.

8.2 Forest management to realize multiple-use forests?

To meet the growing demands on forests to provide diverse ecosystem services (associated with the widespread 'goal-inflation' described in the introduction), it is argued that greater variation in forest management practice is required. In Sweden, the clear-cutting forestry introduced on large scale in the 1950s has been complemented since the 1990s with green tree retention and dead wood preservation to enhance biodiversity (Simonsson 2016). A clear-cutting regime is regarded as superior from a productivity perspective, but it has been heavily criticized by various environmental and social NGOs because of claims, for instance, that this one-size-fits-all solution poses a serious threat to the biodiversity and multi-use potential of the forest landscape (Moen *et al.* 2014). The argument that diversification of management would lead to diversification of forest ecosystem services has also been advanced in forest science literature (e.g. Puettmann *et al.* 2009) and recently incorporated in Swedish national forest policy. A stated policy goal is, for instance, to increase use of continuous cover and mixed species forestry to enhance the multi-use potential of the forest landscape (www.miljomal.se). This has several major perceived advantages. Developing the capacity to apply a greater repertoire of forest management practices (including single-tree selection practices, mixed species forest stands and more diverse tree species in regenerations) should increase the responsiveness of management regimes to forest policy initiatives. Moreover, multiple-use forest landscapes should emanate from diversification of forest management practices, and (hence) facilitate efforts to meet the increasing numbers of goals.

In parallel with the efforts to diversify forest management and increase capacities to meet multiple forest goals, there have been more frequent debates on forest management practices, particularly single-tree selection practices, in the Swedish parliament (Figure 8.1). A significant aspect of this discourse is the alteration of concepts over time. For example, the old Swedish terms for the practice, like 'selective cutting' and 'dimension cutting',

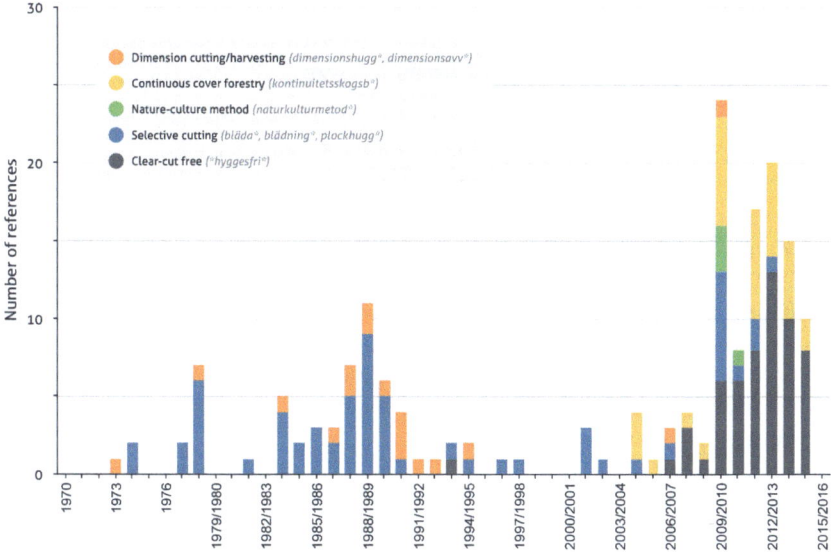

Figure 8.1 Numbers of documents published by the Swedish Parliament (interpellations, motions, bills, chamber protocols, written questions and answers) mentioning terms for single-tree selection practices each year between 1971 and 2015.

Source: The Riksdag's open data website, http://data.riksdagen.se/data/dokument/. Graphical design: Jerker Lokrantz/Azote.

have had very negative connotations. The recent introduction of the vague, and thus more open, concept 'clear-cut free forestry' may have contributed to a broadening of the discussion on alternatives to clear-cutting forestry (Roberge *et al.* forthcoming; Espmark 2017).

From a strictly silvicultural perspective, single-tree selection systems are similar to the clear-cutting system in that they have traditionally prioritized commodity production and viewed other objectives, e.g. maintaining species diversity, as constraints. Today it is often argued that the systems may meet various ecological and social demands on forests better than clear-cutting systems, while there is uncertainty about which system is the most productive (Kuuluvainen *et al.* 2012). Further considerations are that Norway spruce monocultures are most suitable for conversion to stands managed by single-tree selection systems in Sweden, and the proportion of broadleaved tree species in the forest landscape, which is important for biodiversity (Felton *et al.* 2010), may be reduced if single-tree selection systems are used on a large scale.

Due to the strong clear-cutting tradition in Sweden, most forests are now dominated by even-aged stands, with even-age class distributions across the landscapes. Thus, there are now very few suitable full-layered stands for traditional continuous cover forestry. The conversion of single-layered

Box 8.1 Examples of alternatives to standard clear-cutting management practices in Swedish forests

Single-tree selection, or continuous cover forestry, is the most traditional practice that completely avoids clear-cutting. There is an ongoing scientific debate about whether it provides significantly lower or similar productivity to the clear-cutting system (Kuuluvainen *et al.* 2012). However, it may promote a higher abundance of plant and insect species with continuity requirements in forest landscapes, and enhance forests' social and recreational values (Nordström *et al.* 2013).

The creation of chequered gap shelterwoods avoids large clear-cuts. Instead, c. 0.5 ha patches are cut in chequered patterns in forests. When starting the system, the first cutting is done before the normal clear-cutting age and half of the area is cut. The time between the first and second cuts should be half the normal rotation age, so future cuttings can be done at the normal age for clear-cutting. Simulations indicate that this system may result in slightly lower total productivity than a standard clear-cutting system, but increase forests' social and recreational values (J. Sonesson, personal communication).

Prolonged rotations, that is, prolonging the time forest stands remain in a mature stage. This may create forest landscapes that are more attractive for recreational purposes, and enhance both biodiversity and carbon stocks in the forest, but reduce mean biomass production over time and hence profitability (Roberge *et al.* 2016). It may also increase risks of forest damage due to wind-throw and forest pests.

Various forms of shelterwoods can be used to enhance regeneration and avoid the creation of clear-cuts by partly retaining the forest canopy until the new stand has reached a height of about 5–6 m. Retaining shelterwoods can contribute to natural regeneration from seeding (Kuuluvianen and Ylläsjärvi 2011) and provide shelter for planted seedlings (Klang and Ekö 1999), but may result in weaker growth of the regenerated seedlings compared to standard clear-cutting (Bose *et al.* 2014; Agestam *et al.* 1998). Another type of shelterwood system is the natural regeneration or planting of Norway spruce under a shelter of birch (Tham 1994) (Figure 8.4).

Different forms of mixed species forestry increase the number of ecosystem services that can be delivered from the same forest (Gamfeldt *et al.* 2013). Establishment of mixed species forests requires careful planning in the regeneration phase to promote all desired species equally (e.g. Holmström *et al.* 2016; Keskitalo *et al.* 2016). However, levels of each of the multiple ecosystem services provided by multi-species forestry are moderate, and monocultures may provide higher levels of specific services, according to Van der Plas *et al.* (2016).

forests to multi-layered stands is time-consuming and costly due to productivity losses during the transition time (Drössler *et al.* 2014). The basic approach when converting such a homogeneous stand to a heterogeneous stand is to exploit the variation in tree sizes and tree ages that is already present in the stand and to complement it with ingrowth of new, naturally regenerated trees of desired species. Thus, some trees of all diameter classes are cut (and some are retained) in the first thinning. To promote natural regeneration, the retained basal area after thinning must be relatively low.

To avoid this rather demanding process of converting a forest stand initially cultivated for clear-cutting into a continuous cover stand, there are also other clear-cut free alternatives, all of which have pros and cons (Box 8.1). Some of these, like various kinds of shelterwood systems, have been quite extensively tested, as they may facilitate regeneration of dense Scots pine stands for production of high quality timber (Agestam *et al.* 2002) (Figure 8.2). Others, like the chequered gap shelterwood system, have been mainly applied at experimental scale to date (Erefur *et al.* 2011) (Figure 8.3). An advantage of such systems is that light conditions in the extensive edges between stands and clear-cuts suit many plant species, like wild-berry producing dwarf-shrubs, thereby strengthening forests' social values. Clearly, there are well-known alternatives to the dominant clear-cutting practices. However, most of them are apparently associated with lower productivity,

Figure 8.2 A Scots pine stand regenerating from seeding.
Photo: Bo Göran Backström/SKOGENbild.

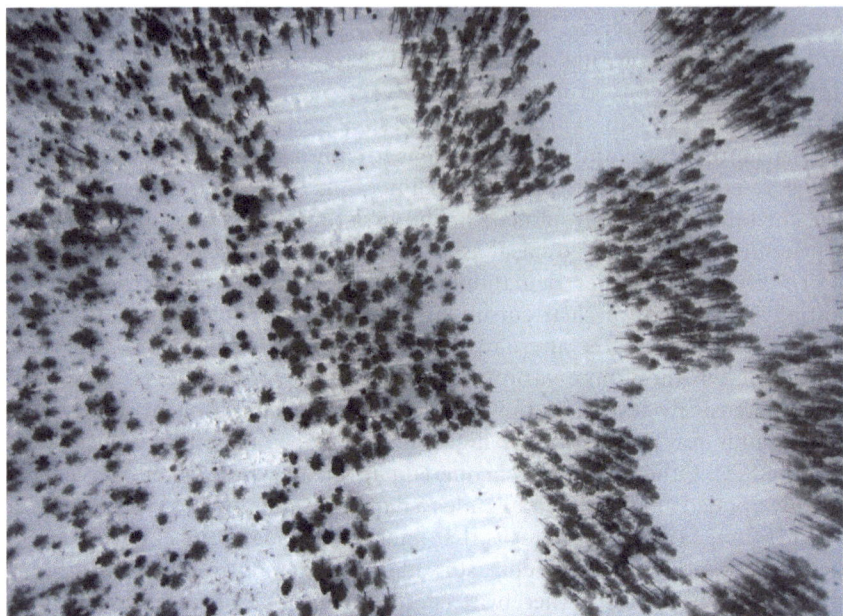

Figure 8.3 A chequered gap shelterwood, with 0.5 ha patches, from above.
Photo: Olle Hagner.

Figure 8.4 A spruce stand regenerating beneath a shelter of birch.
Photo: Lars Rytter.

higher labour costs, and less certain regeneration. Thus, prioritization and trade-offs must be handled in a transparent process in which forest policy goals are closely connected to desired management outcomes.

8.3 Adaptive forest management

In efforts to create opportunities and new ways to initiate and manage change in forestry in a manner that integrates policy and practice better than current processes, the Swedish government commissioned the Swedish Forest Agency and Swedish University of Agricultural Sciences to test an adaptive forest management model. The twin goals of the project (involving close interaction between policy, practice, and research) were to identify processes and practices that could increase biomass production and improve the environmental status of Swedish forests. The resulting model can be regarded as a combination of knowledge-based Adaptive Management and as a strategy to facilitate structured decision-making. It has extended traditional investigations by focusing on stakeholders' needs, motives, and relations, and established a number of trials (Skogsstyrelsen 2016).

As we have seen, Adaptive Management was developed to reduce uncertainties involved in management by experimentation, monitoring, and assessment of outcomes (Holling 1978; Walters 2007; Rist *et al.* 2013). Despite efforts to enable stakeholder participation, Adaptive Management is mainly knowledge-oriented and expert-driven. Moreover, it is a bottom-up and place-specific approach (Holling and Meffe 1996), which may hinder its application on wide spatial scales and integration in a national institutional setting (Rist *et al.* 2016). Consequently, Swedish efforts to apply adaptive forest management differ in several respects from the established definition of Adaptive Management. First, while the Swedish efforts focus on promoting change, the objective of Adaptive Management is to control and avoid unintended change. Second, the Swedish endeavour was a national top-down action initiated by the government and implemented by state authorities, in contrast to the usual local and bottom-up processes of Adaptive Management. Third, in the Swedish efforts stakeholder participation played a key role, while in knowledge-driven Adaptive Management participation is more complementary.

An initial evaluation of the application of the adaptive forest management model indicates that it works (Skogsstyrelsen 2016). Nevertheless, it probably requires tailored development, depending on objectives of the management process; for example, whether it is intended to be a reactive means to reduce uncertainty, a more proactive means to achieve transitional change, or perhaps a combination of both. So far, the adaptive forest management model trialled in Sweden has had much less in common with Adaptive Management than with Transitional Management. This is because both the trialled model and Transition Management are processes intended to initiate strategic change by establishing a transition arena and developing transition pathways and agendas, in a more top-down and participatory manner (Loorbach and Rotmans 2006). However, since the term Transition Management is not mentioned at all in the report from the adaptive forest management project, there is considerable scope to clarify and develop these approaches in Swedish forest management.

Box 8.2 Brief summaries of approaches to the management process in adaptive and transitional management

Adaptive Management is a procedure of planned experimentation to learn about uncertainties in the dynamics of complex natural systems and effects of management practices on these dynamics (Holling 1978; Walters and Hilborn 1978; Walters 2007). It is a structured process, which starts with assessment of the problem(s) and clarification of the management objective(s). Then current knowledge is compiled, and key uncertainties and management actions are identified. The last steps are implementation of management through comparative experiments in the field, monitoring outcomes and evaluation of outcomes, which closes a feedback cycle (Rist *et al*. 2013). Adaptive Management originates from natural sciences and is expert- and research-oriented, but often includes stakeholder participation to increase the chances of successful natural resource management.

In circumstances when dominant structures in society are under pressure from external drivers – for example, climate change – and internal drivers, such as changing values, Transition Management offers practical operationalization to facilitate and accelerate change (Kemp and Loorbach 2006). The key tasks are to establish a 'transition arena' based on expert deliberation to facilitate interaction, knowledge exchange, and learning between and by diverse actors (Kemp and Loorbach 2006), and then develop a transition agenda with a 'basket of objectives' informed by the participants' visions. The next task is to create a broad portfolio of transition experiments, based on the transition agenda, designed to provide results that can be scaled up and potentially contribute to innovation, learning, and (thus) transition toward the defined goals. Monitoring the experiments to provide feedback and learning is also important. This approach originates from socio-technological studies, and collaboration is an inherent feature, but the research is an important expert-guided process intended to obtain valuable information about management alternatives and consequences.

8.4 Dialogue processes and target images

In this book, a central argument is that the gap between governance of forest resources and management on-the-ground has received insufficient attention. Governments may use various policy instruments to address such gaps, including, for instance, law enforcement, subsidies, and voluntary initiatives. During the last two years, the Swedish government has relied heavily on inclusive deliberative stakeholder processes in the form of a National Forest Program (NFP) as part of efforts to meet National Environmental Quality Objectives (Gov. Prop., 2013/14:141). The Swedish NFP should foster long-lasting collaboration and enhance agreement among stakeholders (particularly ENGOs and the forest industry), while generating effective suggestions for the transition towards a bio-based, low-carbon economy (Gov. Bill, 2013/14:141). This is an intriguing case, partly because most European countries have adopted formal NFP processes, and Sweden has been a highly significant exception (Howlett and Rayner, 2006). Due to the specific property rights structure in Sweden with many private forest owners, the Swedish forest governance model has largely relied on the provision of information and advice, through various instruments directed towards private forest owners, but also forest certification schemes, to achieve the goals that would generally be expected from an NFP (Appelstrand 2007).

In terms of organization, the Swedish NFP has an advisory board with approximately twenty members (representing the forest industry, civil society, academia, and public agencies), and several operative working groups. The latter were established during the spring and autumn of 2015 by the Ministry of Enterprise and Innovation, thus they have only recently delivered strategic recommendations for a program. These recommendations are currently being processed in the government offices, and it is still unclear how they will be incorporated in Swedish forest policy and management (Johansson 2016).

In addition to this large-scale governmental process, in recent years the Swedish Forest Agency has initiated a number of parallel stakeholder processes. However, the initiatives have generally focused more on management-level procedures, and set relatively concrete objectives regarding specific issues, such as nature conservation, water-related issues, and recreation. They have been recognized as rather successful, at least from a process perspective. However, it remains to be seen how much these types of processes may contribute to goal achievement and resolution of disputes among stakeholders in the mid- to long-term (Mårald *et al.* 2015; Johansson 2016).

8.5 Concluding remarks

Although a number of integrated approaches have been developed in parallel to ideas about sustainable forest management, none of them has been explicitly applied in a Swedish forestry context. There are several possible reasons for this. The most important may be related to the shifts in forest stakeholders' relative power and roles, common goals, mutual relations, trust, and long-term rules manifested in the three social contracts outlined in Part I. While the second contract essentially followed from the first, the third contract marked a clear shift in policy direction, power relations among involved actors, and governance of the Swedish forests. This shift, which resulted in a strong focus on equal goals, New Public Management, and steering by objectives, entailed a disjunction between governance and management.

Consequently this shift, which included a strong focus on equal goals also tended to, with its strong focus on New Public Management and steering by objectives, entail a disintegration between governance and management (Appelstrand 2012). Several of the approaches reviewed above attempt to integrate governance and management. However, none of them seems to have provided enough guidance in development of the Swedish forestry model. This is likely at least partly because overarching goals of the different approaches reviewed above (such as ecological integrity, resilience and robustness) are not entirely compatible with those of the Swedish forestry model.

Due to the disjunction between governance and management (which has resulted in a strong focus on management and forestry practices, and much less attention to the development of policy and politics) there is also no systematic consideration of how learning could or should feed back to overarching policy levels, the scientific community or established institutions (e.g. parliament). At first glance, this de-politicization may be appealing for policy-makers as it suggests possibilities to foster consensus and problem-solving coalitions across established lines of political conflict (Smith and Kern 2009). However, there are high risks that a learning approach that does not adequately take into account conflicts, power, and tactics will be hijacked by strong interest groups rather than being controlled by groups or individuals engaged in collective learning to promote sustainability. Forest management practices that do not consider the policy and political contexts risk 'problems of fit' with existing political patterns and inaccurate implementation (Meadowcroft 2009). Thus, there is a need to develop an approach with the capacity to integrate governance and management. In the following chapter we argue that Reflexive Forestry may help to do so.

Part IV

Reflexive Forestry

Introduction

The main argument in this book is that there is generally a disjunction between forest governance and management. While a plethora of ambitious, progressive global-, regional- and national-level goals are formulated in our modern 'goal society', several problems hinder their fulfilment. Inherent goal conflicts are often neglected; in a deregulated policy context based on voluntary commitments and market solutions it is very difficult to achieve all aims, and the goal-oriented top-down approach with predefined targets and solutions often meets resistance on a local level and among practitioners. Moreover, although forest science researchers often aim to inform policy and develop practical applications, there is great scope to improve collaborations with actors to incorporate other ways of knowing and make the scientific contributions relevant and useful in different contexts.

A major aim of Reflexive Forestry, the focus of this chapter, is to address these challenges to improve integration of forest governance and management. Another aim of Reflexive Forestry, as outlined in the introduction, is to unite a reflective and analytical tradition of 'questioning and diagnosing' with a positivistic and action-oriented tradition of 'knowing and doing'. Procedures to combine 'opening up' and 'closing down' strategies at appropriate points in social problem-oriented processes are also required. To explore and bring together these aspects of Reflexive Forestry – reflexivity and opening up, and decision-making and closing down – we have focused on three key themes: transtemporal thinking, ways of knowing, and ways of doing. In addition, we have approached the forest as a part of society – a forest arena – where actors engaged in forests and forestry at various levels have been able to establish social contracts to handle common challenges. For some time, such broad, implicit, and explicit social agreements may maintain common understandings of main problems and opportunities, direction of handling, ways of knowing, and ways of doing (cf. Muldoon 2016). However, social contracts may collapse during times of large transitions such as now, when we are facing many uncertainties, and Reflexive Forestry may help efforts to broaden perspectives and find fruitful new directions by bringing forest governance and management closer together.

The purpose of this final part of the book is to outline Reflexive Forestry, building on findings and arguments presented in other parts of the book, concerning the past, future and present (see Figure 2, Introduction). In summary, our exploration of the past, future and present has highlighted an urgent need to develop an approach that allows institutions and practices to respond to changing social-ecological conditions, while closing the identified gap between governance and management. A number of prominent approaches based on ecological integrity (CBD), resilience (Walker *et al.* 2006; Rist *et al.* 2013), and adaptive governance and management (Rist *et al* 2014) have attempted to address these needs. Based on our findings, we are convinced that a reflexive approach is more appropriate for taking these needs into consideration than the alternatives – especially in the context of pressing challenges. In a similar vein to Dryzek and Pickering (2017), we argue that a reflexive approach provides capacity for the flexible reconfiguration required for successful responses to reflection on its performance: 'If institutions are performing poorly, then they need to be able to question their own foundations–rather than just modify their practices while maintaining the same overall identity' (p. 353). As highlighted in section 7.8, a reflexive approach is not without critics. We recognize the weaknesses they have highlighted, and strive to incorporate more refined and robust elements to address potential problems associated with power and democracy. In the following chapter we present and discuss the basic principles of Reflexive Forestry.

However, first it is important to point out what Reflexive Forestry is *not*. It is not a panacea. As emphasized throughout this book, forest governance and management must be adapted to specific contexts. Furthermore, this is not a step-by-step manual. Instead we develop basic principles for Reflexive Forestry, based on findings regarding other approaches to natural resource management with similar aims. However, Reflexive Forestry should be applicable. Thus, we discuss different methods to conduct Reflexive Forestry in practice, but our examples are not a comprehensive list of approaches and tools. Briefly, Reflexive Forestry is a certain way of thinking and a flexible approach for acting when trying to identify and guide the forest arena towards sustainable options in a reflexive and deliberate manner. The overriding goal is to enhance society's ability to sustainably use, preserve, and tend forest resources and environments, by developing knowledge and tools, empowering actors, creating mutual relations, and establishing common understandings. Such efforts and processes may result in a new forest social contract to handle our time's key problems and opportunities.

9 The principles of Reflexive Forestry

9.1 Introduction

The principles of Reflexive Forestry are its fundamental tenets and guidelines for the required formulation of normative objectives, transtemporal thinking, ways of knowing, ways of doing, actors' roles, and generation of directionality. Collectively they provide the foundations for thinking and acting reflexively when maintaining and developing an essential natural and social resource such as the forest. We assume that the forest arena must be both retained and enhanced to maintain vital ecological functions and social relations, adapt to changing conditions, and develop abilities to act in desired directions. Moreover, we acknowledge that individual actions must be connected to common action to enable the arena to handle pressing current challenges, so shared capacity with mutual commitments and pathways that provide betterment for people as well as for society as a whole is required. Accordingly, as a normative point of departure the first principle of Reflexive Forestry is that:

Reflexive integration of governance and management will enhance society's collective capacity to maintain and develop forests as vital and sustainable resources and environments for the benefit of people.

9.2 Transtemporal thinking

Transtemporal thinking, as defined in this book, recognizes that the temporal dimensions of history and future are attached to the present, constantly affecting our understandings, choices, and decision-making in real time (cf. Adam 2013; Hartog 2015; Fareld 2016). Furthermore, we advocate a reflexive understanding of transitional change, acknowledging that the historical evolution of forests and forestry as a social arena is multi-layered, that the future of this social sphere is not given, and that there are many overlapping, legitimate, and competing claims on the benefits of forest resources in the present. Thus, we emphasize that reflexive, well-founded, and responsible decision-making and actions in the present are the last steps in broader efforts to understand forests and forestry in their transtemporal contexts. Transtemporal thinking fosters an ability to shift focus between situations at different times, and connect them to specific present situations and ourselves, thereby inducing both self-confrontation and reflection.

For instance, transtemporal thinking can be used to analyze the six integrated approaches discussed in Part III. Three of the approaches are essentially future-oriented, but with a clear conservative point of departure. The normative objectives of these approaches are to maintain ecological integrity (Ecosystem Based Management/Ecosystem Management), resilience (Resilience Thinking), or robustness (Social Ecological System Framework). Thus, their main aims are to avoid change, minimize future uncertainties and maintain, or even restore past, fundamental ecological and social structures. In contrast, the Ecosystem Service Approach has a presentistic approach but with an implicit future orientation. It holds that ecosystem service assessments should automatically provide guidance towards sustainable use by considering all ecosystem services, benefits, and landscape attributes, and establishing robust methodology to handle them, thus providing a technical and liberal market-based solution to a complex situation. Finally, Reflexive Governance has an ambiguous but still progressive and future-oriented agenda as a normative objective. Sustainable development is seen as a process of social change, where the main problem is to overcome established and destructive 'regimes' while initiating and facilitating transitions in desired directions. Thus, Reflexive Governance is reformistic and goal-oriented.

Reflexive Forestry has wider aims than all of these other integrative approaches, encompassing combinations of conservative maintenance, reformistic change, and inclusion, while establishing methodologies to handle challenges and opportunities. A key issue is when it is appropriate to maintain existing states and arrangements or to promote change. To address this issue transtemporal thinking is valuable, as it enables analysis and establishment of deeper understanding, and identification of salient features that might be missed if transient present conditions are considered in isolation. This enables description of overarching preconditions, path dependencies, dominant perspectives and relations, ongoing transitions, future desires, coming threats and opportunities, and deadlocks. Hence, transtemporal thinking helps efforts to focus reflexively on key current issues, and to highlight resources, services, and processes that are important to maintain and circumstances where change is needed. Consequently, the second principle of Reflexive Forestry is that:

Reflexive and responsible decision-making in the present, constantly informed by transtemporal integration of the past and the future, will enhance actors' capacity to expose and address problems and opportunities.

9.3 Ways of knowing

In order to explore and tackle complex issues such as those associated with the forest arena, different scientific ways of knowing (Pickstone 2001) – world reading, sorting, analysis, synthesis, experimentalism, and techno-science – are important. It is even more important to amplify these efforts with interdisciplinary research to examine connections and establish broader and versatile understandings (with transtemporal perspectives, as outlined above).

Moreover, scientific knowledge of the natural world is not the only valuable kind. Societies all over the world have developed ways of knowing about their environments, referred to as traditional ecological, indigenous, or local knowledge (Berkes 1999; Folke 2004; Rist *et al.* 2010). This knowledge has provided foundations for myriads of activities that sustain societies in many parts of the world and can substantially improve efforts to enhance sustainability. Thus, we promote the establishment of networks with diverse actors and development of interactive and participatory methods to recognize and integrate different ways of knowing when addressing urgent issues. Understanding that people and institutions are unlikely to engage in collaborative action that may threaten their vested interests is also important. Therefore, it also requires attention to power relations and the variations in positions – including scientific positions – held by different actors.

Thus, such integration is far from straightforward. According to Dryzek and Pickering (2017), it is crucial to establish boundary positions between expert and participatory deliberation to foster exchange of knowledge and bridge the gap between theory and practice. Forestry scientists have always acted in such boundary positions (as discussed in Part I), often in collaboration with dominating actors, but also with forest practitioners and small-scale forest owners. Such boundary positions have involved both the tendency of forest researchers to pursue practical goals (embracing political and economic values) as well as fundamental goals (embracing purely scientific values). The former, which often involves practitioners in applied research, is an important way of developing the ability to question pre-suppositions and associated limitations in (for example) forest sciences. The latter is important since only experts can address certain problems (e.g. potential effects of climate change on pest and pathogen populations).

In Reflexive Forestry there is a clear need for analyses with links to the social sphere and practice to inform the normative base. Research should involve interaction with political actors, regarding prioritized issues and implementation of democratically reached decisions, as well as deliberation with diverse actors and stakeholders, to generate useful, embedded, and beneficial outcomes. Similarly, fundamental interdisciplinary and trans-disciplinary research is required in Reflexive Forestry to establish detailed scientific knowledge, develop deep theoretical understandings, and identify

Figure 9.1 Forest excursion with stakeholders and researchers in Finspång, Östergötland 2012.

Photo: Mats Hannertz. Copyright: The Swedish University of Agricultural Sciences.

knowledge-based alternatives relevant to considered ranges of socio-ecological contexts and spatial decision levels.

However, occupation of such a boundary position by scientists requires perspicaciousness and transparency. The establishment of objectives for regulating and managing the forest arena is essentially a political issue open to constant interrogation and re-interpretation. Nevertheless, the strong scientific focus on forest issues has entailed a scientization of policy-making, blurring the boundary between facts and values, and opening possibilities to frame dilemmas in myriads of ways (Pielke 2007). Similarly, as Sarewitz (2004) argues, the 'excess of objectivity', when the underlying value questions are suppressed, has led to deadlocks and conflicts. Instead, with a better understanding not only of facts, conditions, and changes, but also values, positions, and desires, it may become possible to openly and reflexively combine these aspects and develop alternative governing and management options and criteria. This should also enable the assessment and validation of proposed alternatives from both scientific and political perspectives. Hence, the third principle of Reflexive Forestry is that:

Integrating actors' different ways of knowing, by establishing transparent boundary positions, will enable development and exchange of knowledge, and collaborative learning.

9.4 Ways of doing

Using the concept 'ways of doing' enables modes of governance, forest management, and everyday practice to be connected. Reflexive Forestry recognizes a need to unite these three components of 'doing', as they provide complementary strengths to integrate forest governance and management, and openly address and close pressing issues with tangible outcomes.

Recently developed modes of governance often include deliberative processes involving politicians, stakeholders, experts and citizens (Dryzek 2000; Bäckstrand *et al.* 2010). Moreover, there are already clear signs that the role of the Forest Agency in Sweden is about to change in this direction. To facilitate such processes on the national level National Forest Programmes have been established in Sweden, and many other countries worldwide (Saarikoski *et al.* 2010; Winkel and Sotirov 2011; Larsson *et al.* 2014; Johansson 2016).

These efforts to handle changing situations reflect a more general transition 'from government to governance' (Bevir 2013). This shift of steering emphasizes the plurality of interrelated policy arenas, the mutual exchange of knowledge and negotiated interactions between different actors, and the blurring between public and private spheres of society (Ansell and Gash 2008). Moreover, it involves transformation of the link between the state and its national territory, as increasing numbers of policy decisions are made by international and transnational organizations, but also by regional and local governments. According to Torfing *et al.* (2012) this shift seems to have produced a number of irreversible changes. These include, for instance, expectations that individual and collectively organized stakeholders should become actively involved in policy-making, the transformation of public agencies to meet these changes in efforts to solve problems jointly, and the perception that various forms of interactive governance modes are considered legitimate alternatives to traditional hierarchical governance modes. These changes have often been accompanied by development of public–private partnerships and other decentralising instruments, but sometimes without decentralization of resources, reflecting a different role of the state (Bjärstig and Sandström 2016).

It is important to note that this shift does not necessarily involve 'hollowing-out' of the state, but rather a change in the state's role. According to Meadowcroft (2007), this steering logic involves an important role for public authorities at all levels, where 'government' is central to 'governance for sustainable development'. Thus, governments will continue to play crucial roles in governance and management within their respective jurisdictions, both as a central and resourceful participant and as a meta-governor that facilitates, manages, and directs various governance arenas, and bridges the gap we have identified between governance and management. It is also important to be aware that shifts in governance entail shifts in power relations, potentially raising risks of interest group capture (and no guarantee that participation will provide the ability to influence decisions), 'green-washing' and/or 'rubber stamping' of decisions favouring the most powerful actors. Thus, attempts to adjust governance modes to meet given objectives must include critical appraisal of power relations and risks of interest group capture, together with careful development of processes to avoid side-lining any relevant perspectives when addressing issues.

The organization of this way of doing fits well with what Denhart and Denhart (2007) defined in terms of New Public Service. In contrast to Old or Traditional Public Administration and New Public Management, in this steering mode the role of the public servant is to open up a social arena by enabling citizens to articulate and meet their shared interests and enable common action. Good order, efficiency, and productivity (the main objectives of Old and New Public Management) are still important. However, democratic values are paramount. The focus is on negotiating and brokering interests among actors (citizens, not consumers) and building coalitions between the public and private sectors, and between different groups and organizations (Denhardt and Denhardt 2007). In this process the government's role is transformed from prescribing and controlling to agenda setting, bringing various actors to the table, and brokering efforts to clarify desired directions, what needs to be done, and by whom, then put the solutions into practice (Denhardt and Denhardt 2007; Meadowcroft 2007; Appelstrand 2012).

Similarly, Reflexive Forestry promotes a way of doing that facilitates collaborative formation and closing of shared understandings, resolutions, responsibilities, and collective action to guide the forest arena towards sustainable development. Instead of launching ambitious goals with pre-set targets and expert solutions from above, politicians, scientists, and other actors initiate dialogue and action by openly addressing major dilemmas and opportunities that require urgent attention. This creates space for individuals, groups, companies, and other organizations to engage and become important contributors and developers of context-relevant ambitions, strategies, resolutions, and practices that are collectively capable of tackling the overarching issues in diverse ways (Stilgoe *et al.* 2013). In this context, the state has a pivotal role to 'provide' actors with a common arena for dialogue processes, interaction, and development. The state also has the resources and ability to fund, guide, empower, level inequalities, and sustain different actors and activities in an integrated manner, with long-term commitment, to make the arena genuinely inclusive and open.

Moreover, since different groups of actors' conceptions of the world differ, their framings of the 'object' of governance and its boundaries will differ. How these different framings are interactively and mutually negotiated has an important bearing in reflexive governance. Consequently, the fourth principle of Reflexive forestry could be summarized as follows:

To integrate forest governance, management, and practice, it is necessary to couple different levels, sectors and actors, openly addressing challenges, building capacity, and enabling the creative engagement of actors.

9.5 Concluding remarks

The principles of Reflexive Forestry constitute guidelines for a way of thinking and acting reflexively when maintaining and amending the forest arena (Box 9.1). Transtemporal thinking is crucial to promote reflexivity, to enable identification of salient and urgent issues, and to highlight important aspects to maintain or (when necessary) change. Moreover, by building networks with a multitude of actors and developing interactive and participatory methods, it is possible to link different ways of knowing and theory and practice. Regarding ways of doing, Reflexive Forestry sees forest governance, management, and everyday practices as tightly interlinked. To exploit this linkage as fully as possible, it is important to consider pressing issues openly and inclusively, then handle them by releasing the potential in society, from above and below, to enable reflexive, innovative, diverse, and tailored forest management.

Finally, a problem when addressing interconnected forest issues is the fragmentation of policy-making, with separate institutional settings, involved professionals, stakeholders, and contextual understandings. To contribute

---

Box 9.1 Principles underpinning Reflexive Forestry

1   Normative objective: *Reflexive integration of governance and management will enhance society's collective capacity to maintain and develop forests as vital and sustainable resources and environments for the benefit of people.*

2   Transtemporal thinking: *Reflexive and responsible decision-making in the present, constantly informed by transtemporal integration of the past and the future, will enhance actors' capacity to expose and address problems and opportunities.*

3   Ways of knowing: *Integrating actors' different ways of knowing, by establishing transparent boundary positions, will enable development and exchange of knowledge, and collaborative learning.*

4   Ways of doing: *To integrate forest governance, management, and practice, it is necessary to couple different levels, sectors, and actors, openly addressing challenges, building capacity, and enabling the creative engagement of actors.*

5   Forest social contract: *To develop a forest social arena, it is necessary to agree upon common sets of rules and understandings that integrate all forest related issues and clarify associated actors' rights and obligations.*

6   Methods to gain directionality (toolbox): *Transtemporal studies, collaborative and reflexive processes, tailored forest management, and feedback and learning (see Chapter 10).*

to efforts to counter this fragmentation, we have used the term 'forest arena' instead of the usual 'forest sector', which reinforces the compartmentalization. This arena is composed of the forest resources, benefits, rights, and obligations, together with the multitude of associated actors, organizations, institutions, and markets, with their various ways of knowing, ways of doing, and motivations. The term 'arena' also indicates that there are competitive claims, coalitions, and interactions among the actors. However, as in every arena, there are common rules for coexistence, including both explicit regulations and implicit norms, which constantly formulate and reformulate what is seen as acceptable and unacceptable, and desirable and undesirable behaviours and ambitions.

To achieve constructive change and collective action the aim of Reflexive Forestry is to establish preconditions and mutual trust, so that (as far as possible) all these actors can interact under a common set of evolving rules – a forest social contract – that promotes a common directionality and action. Accordingly, the fifth principle is that:

To develop a forest social arena, it is necessary to agree upon common sets of rules and understandings that integrate all forest related issues and clarify associated actors' rights and obligations.

10 The toolbox of Reflexive Forestry

10.1 Introduction

The toolbox is a set of methods, approaches, and other tools that can be used in various ways to meet the objectives of Reflexive Forestry, following the principles set out above. It includes technical solutions, ecological measurements, and applications *in* the forest or associated technological systems such as those typically used in applied forest sciences. However, since Reflexive Forestry is intended to engage the entire forest arena, the toolbox is more focused on approaches to find starting points, enhance mutual interactions, develop context-relevant means, and enable learning and feedback. As already mentioned, we provide a few examples, but have no intention of listing all tools that could be useful. This is partly because the list would be extremely long (and continuously growing), and presenting it would be pointless. More importantly, Reflexive Forestry adopts a pragmatic attitude to measurements, practices, management systems, and frameworks, recognizing them all as potentially valuable, as long as they are used and combined in a reflexive, inclusive, and responsive way to initiate constructive processes. Thus, together with the guiding principles, the toolbox can enable the development of shared capacities to handle the forest arena.

10.2 Transtemporal studies

Transtemporal studies examine in a coherent way how past and future dimensions feed into current situations. Thus, they embrace two forms of presentism, which is regarded (especially by historians) with great scepticism. This is because of claims that it corrupts historical methodology by tending to interpret the past in presentist terms and shifts historical interest toward the contemporary period (Hunt 2002). However, Oreskes (2013) counters that this indicates confusion between method and motivation. All historical studies are to some degree imbued with 'motivational presentism', where 'what matters to us about the past has everything to do with who we are, where we live, and what we think is important – to us, here, now, in the present' (Oreskes 2013, p. 603). Without deviating from the rigor of historical methodology, and (in the context of this book) methods used in futures studies, acknowledging these motivations allows connections to be drawn between the present and both historical and futurological findings. The connections can then be used to broaden potential audiences, enrich public debate, and set current notions, positions, and politics in deeper perspectives (Sörlin 2011).

Thus, historical studies, as tools in Reflexive Forestry, should be characterized by motivational presentism, approaching present issues by seeking their origins and narratives globally, nationally, or locally. Historical studies can accentuate continuities and change, form contrasts and show alternative ways to think and act, thereby highlighting salient features today (McNeill 2000). Thus, such studies can illuminate present situations in multi-layered and nuanced ways. Moreover, historical case studies can be used to investigate how earlier efforts to promote change in certain directions were conducted. Although history never repeats itself, such case studies can be highly instructive, facilitating efforts to understand current situations, and guide transformative efforts (Geels 2002).

Similarly, futures studies that provide as rich and diverse visions of the future as possible are extremely valuable for broadening understanding and opening up discussion of current options. This requires use of diverse methods, foci, and framed futures, including efforts to combine qualitative and quantitative approaches, projections, modelling, and both exploratory and anticipatory scenario analysis. Moreover, futures studies can facilitate interactions with societal actors, by using established predictions and scenarios to initiate discussions about options and consequences. A more active approach is to conduct futures studies cooperatively with various actors to develop wider arrays of scenarios and implications for forest practice. Likewise, backcasting with stakeholders is an effective way to establish diverging desired futures, and map current positions and values (Sandström *et al.* 2016).

Thus, historical and futures studies, separately or combined, enable transtemporal learning, in several ways. First, transtemporal studies can highlight pressing issues in a reflexive manner, thereby reducing risks of erroneous causal and precipitate connections being drawn between certain interpretations of current situations and specific solutions. Moreover, they can help identify appropriate times to act (which is often difficult with a narrow temporal understanding), by indicating when change is possible or difficult to attain, and when the status quo *must* be challenged.

10.3 Collaborative and reflexive processes

To handle complex challenges and opportunities, it is vital to establish a collective directionality, to address current myopia, anticipate risks and opportunities, and initiate appropriate common action. In representative democracies, this is ideally handled through elections and parliamentary assemblies at various levels. Citizens commission policy-makers to address their concerns and interests, and the policy-makers in turn commission administrators to implement policies and perform identified tasks.

However, with the increasing complexity, uncertainty, and/or ambiguity of problems, representative democracy and its established processes do not seem to be fully capable of managing the associated conflicts, particularly if different interests and value systems are at stake. Accordingly, legitimacy declines since political solutions cannot be achieved.

In the literature and several of the integrated approaches explored in Chapter 7, the cardinal way to handle this problem is through interaction, communication, and dialogue. It has been argued that collaborative governance and management can be used to identify and fruitfully discuss uncertainties, conflicting or divergent goals, benefits and values, and steer the handling of problems (Norton 2005). These types of processes have also been recognized as ways to secure and enhance the legitimacy of policy-making (Ansell and Gash 2008; Lockwood *et al.*, 2010; Emerson *et al.* 2011). It is assumed that actors are more likely to consider decisions legitimate, and support them, if they have participated in a policy-making process, thereby reducing conflicts and resistance. Thus, enhancing legitimacy by promoting collaborative policy-making between stakeholders with interests in the management and use of forest resources is likely to enhance the effectiveness of forest policy implementation (Reed 2008; Raitio and Harkki 2014).

Of course, from a government perspective it is impossible to include actors in every decision-making process. Inspired by a reflexive governance approach we therefore suggest that it is necessary to establish processes and institutional arrangements that systematically address complexity, uncertainty, and ambiguity based on a division of roles and labour between different levels and actors, as outlined in Table 10.1. (Klinke 2009). In routine situations, where issues do not involve high degrees of complexity, scientific uncertainty, and/or socio-political ambiguity (or require any policy or institutional changes), a traditional top-down or state regulation approach is adequate to ensure collective directionality. However, in cases with increasing complexity, where there is potential for conflict associated with different ways of knowing, it will be useful to involve scientists or other experts with policy-relevant knowledge to collaboratively share, evaluate, and exchange information and knowledge in recognized advisory expert bodies. The primary aim of these bodies will be to develop relevant policy alternatives, since they do not have decision-making power. On a global level, the IPCC and IPBES play such advisory roles. In a similar vein, advisory expert bodies may be appointed at national, regional, and even local levels.

Issues associated with high uncertainty often require a more complex evaluation procedure. Expert knowledge is still valuable as a basis for handling conflicts, but the collected information has to be brought into a deliberative arena with actors that are directly affected by the issues, at multiple levels and within multiple sectors. Due to the lack of scientific

Table 10.1 Four categories of collaborative and reflexive policy-making processes, and the actors who could be engaged in them (developed from Klinke 2009)

Character of issue	Simple	Complex	Uncertain	Ambiguous
Ways of doing	State-centred regulation: • Routine operation	Expert deliberation: • Scientific analysis	Stakeholder deliberation: • Scientific analysis and balancing of interests	Inclusive deliberation: • Trade-off analysis and deliberation, • Scientific analysis and balancing of interests
Type of conflict	Not applicable	• Cognitive	• Cognitive • Evaluative	• Cognitive • Evaluative • Normative
Actors	• Policy-makers • Authorities	• Policy-makers • Authorities • Scientific experts	• Policy-makers • Authorities • Scientific experts • Stakeholders	• Policy-makers • Authorities • Scientific experts • Stakeholders and the public

answers to these issues, the primary objectives are to deal with risk and uncertainty, then evaluate, and potentially decide, how much uncertainty and risk the involved actors are willing to accept in order to pursue some future opportunity. These processes (such as the setting of common environmental targets) may be advantageously associated with adaptive or transition management.

Finally, when issues imbued with socio-political ambiguity must be addressed, it is essential to find solutions that are compatible with normative aspects, such as the interests and values of the people affected, and to resolve conflicts among them. Here, collaborative and reflexive processes are required that involve representatives of governments, the public, relevant agencies, and stakeholders. On a national level, examples of these processes include National Forest Programmes, while important regional- and local-level factors may include the systematic involvement of actors in policy development or municipal plans to resolve ambiguities and norm/value conflicts. It is important that the deliberative processes are institutionalized, and not decoupled from the institutions of representative democracy. The aims are to mobilize the knowledge of those affected, take normative as well as value-oriented preferences into consideration, and foster learning and exchange between different actors.

The suggested division of roles based on increasing potential for conflict, along a spectrum from simple, through complex and uncertain to ambiguous (see Table x) will require new ways of doing things that have implications for the roles of public authorities at various levels, including the municipalities. In this book we assume that we are moving away from new public management to new public service. Within the New Public Service framework, public agencies are responsible for setting up expert advisory bodies, or arenas not only for associational policy-making but also collaborative governance foras, and for facilitating outcomes that could not, supposedly, be accomplished solely by the state (Ansell and Gash 2008; Emerson *et al.* 2011). This means that the state may retain control over the agenda and policy processes, while stakeholders also play key roles. However, these types of collaborative arrangements may also develop independently of the state, as we have seen in relation to certification schemes. Typically, collaborative and reflexive processes involve communication and interaction procedures, during which actors develop an understanding of each other's preferences through dialogue and deliberation, which may bridge different ways of knowing and doing. This mutual knowledge may provide important foundations for the shared diagnosis of complex problems, and potentially for collective ways of doing in terms of changing policy, management and everyday practices (Webler *et al.* 1995; Pahl-Wostl 2002; Muro and Jeffrey 2008).

We assume that this division of roles and labour will provide a basis for closing the gap between governance and management. However, the suggested approach is primarily related to policy-making and the governance of forests and related issues. To be able to implement policy into practice it is also necessary to develop and match the management tools to the policy objectives and governance processes. How this can be done is the subject of the next section.

10.4 Tailored forest management

Tailored forest management is a key concept in Reflexive Forestry and is intended to provide customized management solutions for specific socio-ecological contexts and the associated actors at all relevant levels. Its implementation will address both overarching societal challenges and opportunities, and local people's situations and aspirations, with the aim to generate benefits at both levels. Hence, tailored forest management requires the identification of diverse possible management regimes and, following careful consideration, closing down of all but one of them. In this process, some key options must be considered, regarding the management process, management practices, and technology to be applied.

The management processes to choose from are Adaptive Management, Transition Management or a combination of the two. The optimal choice depends on the desired relationship between management and governance. If the governance objective is to maintain conditions in the focal socio-ecological system(s), fill cognitive gaps, or reduce uncertainty, Adaptive Management is suitable. However, if the governance objective is to overcome institutional, strategic, and normative obstacles, or try to find new pathways for development, Transition Management is more powerful. Although these two management processes have many similarities and are not mutually exclusive, Adaptive Management concentrates on reducing uncertainty, while Transition Management focuses on innovations to achieve defined objectives. It should be noted that traditional forest management includes features resembling aspects of both Adaptive and Transitional Management, such as learning by doing, field trials, interaction with practitioners, transition experiments, monitoring, and evaluation. A key step in tailored forest management is to make an informed and deliberate choice between the two management processes, or specified combination of them, due to the assumed differences in their connections to forest governance.

The next choice to be made in tailored forest management is that of forestry practice. Here it is important to acknowledge how previous and current management may constrain the availability of alternatives for the future. For instance, the conversion of even-aged stands initially cultivated for clear-cutting into uneven-aged stands suitable for continuous selective cuttings takes decades of well-planned thinnings that open the forest canopy at appropriate times and places to allow continuous natural regeneration. Similarly, the development of a mixed species forest requires decadal long-term planning of sequential management interventions that carefully structure the inclusion of different species in the forest stand as the canopy slowly closes. The described circumstances highlight the importance of transtemporal thinking when choosing forestry practices; the historical constraints, long lead-times, and uncertainty of future outputs are all connected to, and demand, reflexive and responsible decision-making in the present.

In countries where forests are privately owned, the forest owners are the ultimate decision-makers in terms of management practices (within constraints imposed by practical realities and legislation). As increasing numbers of forest owners live in cities, and thus are increasingly losing touch with local knowledge traditions, there is a growing need for support in forest management decisions. Such decision support not only needs to connect the forest owners to local forestry knowledge, but also to consider available indigenous knowledge (notably the knowledge of Sami reindeer herders in northern Sweden) and scientific knowledge. Ultimately, the decision support should support the merger of these three ways of knowing, enabling forest owners to identify customized ways of doing, or tailored forest management, that translates global and national forest policy into local forestry practices.

The ultimate challenge here is to find ways of doing that may integrate the objectives of the individual forest owner with those of other societal actors, thereby closing the gap between forest governance and management. Interactive forest counselling directed towards forest owners may make an important contribution to the integration of different ways of knowing and doing. However, many societal objectives for forests, such as biodiversity protection, maintenance of water quality, and amelioration of climate change cannot be delivered at the individual stand level and must be addressed at larger spatial scales, e.g. landscape, watershed, or even national levels. Hence, researchers have identified a need to shift planning focus from, for example, the stand level to the landscape level to reconcile ecological and socioeconomic dimensions in land-use planning and the broader sustainable development agenda (Ranius and Roberge 2011; Lundmark *et al.* 2016). This includes needs to adjust planning approaches and incorporate consideration of land-use activities and users (*inter alia*) in comprehensive municipal planning. To what extent this is possible, and the requirements in terms of incentives for forest owners to collaborate in a Swedish context, are still largely unknown (Carlsson 2017; Wallin 2017). However, in order to close the gap between governance and management it is essential to develop effective methods to motivate forest owners to collaborate across large spatial scales. This will inevitably also affect the integration of various policies on, for instance, timber and bioenergy production, water management, and biodiversity conservation at the landscape level.

Sufficiently advanced technology (which is being constantly developed, and hence offering increasingly refined options) is essential for tailored forest management. For example, digitilized hydromapping equipment in forestry machinery can be used to design riparian buffers with variable widths during logging operations (Kuglerová *et al.* 2014; Ågren *et al.* 2015).

With the ability to track the flow of groundwater, drivers of forestry machinery can easily identify sensitive areas and mitigate negative effects of logging and soil perturbation, such as changes in groundwater pathways, and increases in siltation and exports of metals and metal-containing pollutants, e.g. methylmercury (Ågren *et al.* 2014). Thus, riparian buffers of variable widths, with wider retained forests in wet riparian hotspots and narrower buffers in less sensitive areas, can enhance ecosystem services such as water quality and biodiversity, without necessarily incurring wood production costs. Similar technological advances in numerous fields are enabling increasingly efficient and gentle forestry practices, and hence increasingly well-tailored forestry management. The challenge for forest owners is to maintain awareness of technological developments and the costs associated with using advanced technologies.

Overall, key steps in tailored forest management are to make deliberate choices between adaptive and transitional management (or a combination of them) and goal-oriented forest management practices, and then identify and deploy efficient technological solutions. This requires interaction and communication among actors and processes at global, national, and local governance levels, not least the local stakeholders who must implement the selected management solutions. By linking challenges, knowledge, technology, and experiments to local engagement and initiatives, a constructive space can be created to foster the emergence of tailored innovations and (after risk- and cost-benefit analysis and negotiation) close specific resolutions with beneficial real world outcomes. Hence, these divergent resolutions can bridge the disjunction between governance and management, and improve responses (in various ways) to general (or local) dilemmas and opportunities.

10.5 Feedback and learning

With numerous tools enabling numerous possible resolutions and outcomes, there is a need in any management system for monitoring, evaluation, and feedback. Here too, the long timescales of forests and forestry present challenges: rigorous empirical evaluation of a change may not be possible for more than fifty years. Thus, a change in management practice is in itself an experiment that cannot be tested in advance. However, as already mentioned, knowledge and experience have accumulated over many years, and an array of tools is available to predict future outcomes. Both Adaptive and Transition Management include feedback loops that can provide early warnings. Moreover, material in long-term field trials, which have been established since the early twentieth century with specific purposes, can be used to address new questions. Because of the characteristically long timescales and multi-actor relationships of the forest arena, state-centred regulation is essential to ensure enduring monitoring and feedback procedures. This state-centred regulation should also entail responsibility to establish inclusive participation and open reporting.

As we have seen, feedback is always included in integrated approaches and management processes, but how it should be processed and channelled back to relevant actors in the arena is not always clarified. Particularly little is known about how to link horizontal governance, such as collaborative processes, consultation procedures, and collective management of forest resources, to vertical governance structures typical of traditional decision-making (Rogers and Weber 2010). Thus, it is important in collaborative and management processes to identify relevant actors and institutions at all levels and layers, and establish channels within and among them to allocate responsibility and achieve tangible outcomes. For example, policy recommendations should be directed and discussed with concerned political players or relevant decision-makers, so as to inform policy deliberations and improve decisions. Similarly, practical and administrative implementation tasks should be discussed with private actors, local groups of actors, involved companies, or authorities to identify appropriate measures, while issues requiring better knowledge could be discussed with researchers, or used to stimulate collaborative processes (Mårald *et al*. 2015).

However, monitoring, assessment, and evaluation are parts of a broader learning process. Together, transtemporal studies, collaborative processes, tailored forest management and feedback comprize an approach to grasp

Figure 10.1 In the summer of 2014 a forest fire raged for two weeks in Västmanland (mid-Sweden) totally destroying c. 14,000 hectares of forests. A stakeholder-researcher excursion in the fire area in autumn 2015.

Photo: Lars Klingström. Copyright: The Swedish University of Agricultural Sciences.

the forest arena and enable social learning (Pahl-Wostl 2006; Ison and Watson 2007: Mostert *et al.* 2007). This is a process of social change shaped by dynamic interaction between people and the forest environment in the construction of meaning and identity (Muro and Jeffrey 2008). This may in turn lead to the acquisition of knowledge, and various types of skills and changes in cognitions and attitudes. It may also enable trust-building and improve social relationships. The resulting social learning may contribute to common understandings and, potentially, mutual agreements and collective action (Muro and Jeffrey 2008). Accordingly, it can generate changes in understanding that extend beyond the individual to become situated within wider social contexts (Reed *et al.* 2010) – a social forest contract.

10.6 Concluding remarks

The Reflexive Forestry toolbox includes: transtemporal studies, which can identify key starting points for opening up reflexive consideration; collaborative processes to establish mutual communication and interaction, thereby enhancing mutual understanding and directionality; tailored forest management to coherently address overarching challenges, and local conflicts, situations, and aspirations, then close down customized solutions; and monitoring to channel feedback and enable social learning that embraces wider social units. In the current situation, these tools could be seen as costly and inefficient. However, with a shift of emphasis towards more practice-orientation and New Public Service, working procedures can be redirected so that interaction and collaboration become routine elements of research, policy-making and implementation.

As already stressed, this toolbox is not a panacea, but a set of procedures that enable 'muddling through' (Lindblom 1959), with the normative objectives to find common ground, collective pathways, and shared capacities to handle the forest arena. However, it should be stressed that Reflexive Forestry does not strive for consensus and harmony. The forest arena is competitive, populated with actors with diverse understandings, motivations, and ways of knowing and doing, and there will always be different opinions, attitudes, and interests. This complexity raises numerous obstacles, but the diversity of positions and tensions, if handled inclusively, may also stimulate self-confrontation and reflection, and facilitate common rule-making and development.

This book is a first attempt to establish an approach that can be applied in management systems such as forestry, and we realize that it has various shortcomings. However, this also means that there are numerous opportunities to develop and clarify the identified principles and toolbox. Therefore, we urge colleagues and actors in the forest arena to further develop the Reflexive Forestry approach to increase our capacity to meet future sustainability challenges.

11 Towards a new forest social contract?

The forest arena, with resources, benefits, actors, organizations, and associated ways of knowing, ways of doing and motivations, is a vital part of Swedish society. Despite diverse competing claims and interests, at various times sets of implicit and explicit agreements and game rules have been formed. Such 'forest social contracts' cover what are seen as the main challenges, what are agreed as the best ways of knowing and doing, and what the common direction should be. The first contract arose in response to the rapid industrialization that threatened both sustainable forest use and social conditions, while the second contract arose through perceptions that rationalization of forestry could contribute to development of a welfare state. In both of these cases the state, the forest industry, and private forest owners played key roles in building up shared capacities (institutions for regulation and implementation), the acquisition and dissemination of knowledge, and norm formation. However, when conditions changed, and new knowledge, values, and challenges appeared, the functionality and legitimacy of the contracts weakened.

During the last twenty-five years the third contract – the Nordic or Swedish forestry model – which seeks to balance production and biodiversity by entrepreneurialism, voluntarism, and ambitious goal setting, has prevailed. In this book we have criticized this model for problems with goal inflation, implementation deficits, myopia, and a lack of integrated ways to introduce new forest management tools. We have also pointed out that ongoing changes, such as globalization, energy transitions, and value shifts, have put pressure on the third contract. Notably, it does not yet incorporate any measures or processes to counter or adapt to climate change, which traditional actors use to support existing positions, rather than to stimulate new ways of knowing and doing. However, there are signs that new approaches to climate change have started to reconfigure arguments and practices, involve new actors, and shift actor coalitions.

Hence, building on our findings and arguments, it is possible to trace an ongoing revision of the third contract, or even a shift towards a new, fourth contract. Our reflexive and transtemporal perspective conflicts with drawing precise inferences about this possibility, since it is an open-ended social process. Nevertheless, we argue that Reflexive Forestry, with the presented principles and tools, is a suitable integrated approach to guide such a transition. This would involve identifying and opening up important issues and starting points, establishing communication and interaction, opening a constructive space for evolving a new tailored form of forest management, and social learning to enable a common directionality and shared capacity to make necessary changes to the forest arena (adopting reflexive and inclusive approaches at each stage).

However, regardless of how a forest social contract is constituted, such agreement always means a 'closure', as some actors' goals, understandings, ways of knowing and ways of doing become predominant, while others' become marginalized. This is inevitable and may also be advantageous when collectively tackling crucial dilemmas and opportunities in order to achieve sustainability. Nevertheless, in such closure processes, when a new forest contract is forged, reflexivity is especially important to ensure transparency regarding implemented options and the reasons (normative and scientific) for selecting them. It can also highlight important aspects such as who benefits, for what purpose, and who is hurt?

Moreover, in the long run forest social contracts tend to stagnate and impede society's capacity to change and adapt to ongoing transitions, thereby eroding sustainability. Understandings become obvious; coalitions of involved actors become cosy and informal; and ways of knowing and doing become routine. Thus, even in periods when a specific forest social contract dominates the forest arena, transtemporal thinking and reflexivity are important. This way of thinking and acting stimulates vigilance and attention to changes, anomalies, and what is really important in a well-functioning society.

References

Abson, D.J., von Wehrden, H., Baumgärtner, S., Fischer, J., Hanspach, J., Härdtle, W., Heinrichs, H., Klein, A.M., Lang, D.J., Martens, P., and Walmsley, D. (2014). Ecosystem services as a boundary object for sustainability. *Ecological Economics* 103, 29–37.

Adam, B. (2013). *Time and social theory.* Cambridge: John Wiley & Sons.

Adam, B., and Groves, C. (2007). *Future matters: Action, knowledge, ethics.* Leiden, the Netherlands: Brill.

Agestam, E., Blennow, K., Carlsson, M., Niklasson, M., Nilsson, S. G., Nilsson, U., . . . and Sverdrup, H. (2002). Productivity scenarios for the Asa Forest Park. In *Developing principles and models for sustainable forestry in Sweden*, 355–380. Springer Netherlands.

Ågren, A. M., Lidberg, W., Strömgren, M., Ogilvie, J., and Arp, P. A. (2014). Evaluating digital terrain indices for soil wetness mapping – a Swedish case study. *Hydrology and Earth System Sciences*, *18*(9), 3623–3634.

Ågren, A. M., Lidberg, W., and Ring, E. (2015). Mapping temporal dynamics in a forest stream network – Implications for riparian forest management. *Forests*, *6*(9), 2982–3001.

Allard, C., and Skogvang, S. F. (eds). (2015). *Indigenous Rights in Scandinavia: Autonomous Sami Law.* Farnham, UK: Ashgate Publishing.

Allard, C., Elenius, L., and Sandström, C. (eds). (2016). *Indigenous rights in modern landscapes: Nordic conservation regimes in global context.* New York: Routledge.

Allen, C. R., Fontaine, J. J., Pope, K. L., and Garmestani, A. S. (2011). Adaptive management for a turbulent future. *Journal of Environmental Management*, *92*(5), 1339–1345.

Andersson, J. (2012). The great future debate and the struggle for the world. *The American Historical Review*, *117*(5), 1411–1430.

Andersson, J., and Keizer, A. G. (2014). Governing the future: science, policy and public participation in the construction of the long term in the Netherlands and Sweden. *History and Technology*, *30*(1–2), 104–122.

Ansell, C., and Gash, A. (2008). Collaborative governance in theory and practice. *Journal of Public Administration Research and Theory*, *18*(4), 543–571.

Antonson, H., and Jansson, U. (eds). (2011). *Agriculture and forestry in Sweden since 1900: geographical and historical studies.* Stockholm: Royal Swedish Academy of Agriculture and Forestry.

Appelstrand, M. (2007). *Miljömålet i skogsbruket: Styrning och frivillighet.* Diss. Lund, Sweden: Sociologiska institutionen, Lunds universitet.

Appelstrand, M. (2012). Developments in Swedish forest policy and administration – from a 'policy of restriction' toward a 'policy of cooperation'. *Scandinavian Journal of Forest Research*, 27(2), 186–199.

Arbetsgruppen för framtidsforskning, Justitiedepartementet. (1972). *Att välja framtid: ett underlag för diskussion och överväganden om framtidsstudier i Sverige: betänkande*. Stockholm: Allmänna förl.

Árnason, J. P., and Wittrock, B. (2012). *Nordic paths to modernity*. New York: Berghahn Books.

Arts, B., and Buizer, M. (2009). Forests, discourses, institutions: A discursive-institutional analysis of global forest governance. *Forest Policy and Economics*, 11(5), 340–347.

Asdal, K. (2011). *Politikkens natur: Naturens politikk*. Oslo: Universitetsforlaget.

Åslund, Å. (2008). *Allemansrätten och markutnyttjande – studier av ett rättsinstitut. (The right of public access and land use – studies of a legal institution)*. Diss. Department of Economical and Industrial Development, Studies in Arts and Science, Philosophical faculty, Report, (434). Linköping: Linköping University.

Backman, F., and Mårald, E. (2016). Is there a Nordic Model for the treatment of introduced tree species? A comparison of the use, policy, and debate concerning introduced tree species in the Nordic countries. *Scandinavian Journal of Forest Research*, 31(2), 222–232.

Bäckstrand, K., Khan, J., Kronsell, A., and Lövbrand, E. (eds). (2010). *Environmental politics and deliberative democracy: Examining the promise of new modes of governance*. Cheltenham, UK: Edward Elgar.

Baker, S. (2007). Sustainable development as symbolic commitment: Declaratory politics and the seductive appeal of ecological modernization in the European Union. *Environmental Politics*, 16(2), 297–317.

Baldwin, A. (2003). The Nature of the Boreal Forest: Governmentality and Forest-Nature. *Space and Culture*, 6(4), 415–428.

Baycheva-Merger, T., and Wolfslehner, B. (2016). Evaluating the implementation of the Pan-European Criteria and indicators for sustainable forest management – A SWOT analysis. *Ecological Indicators*, 60, 1192–1199.

Beck, U. (1992). *Risk society: Towards a new modernity*. London: Sage Publications.

Beck, U., Bonss, W., and Lau, C. (2003). The theory of reflexive modernization problematic, hypotheses and research programme. *Theory, Culture & Society*, 20(2), 1–33.

Bell, W. (1997). The purposes of futures studies. *The Futurist*, 31(6), 42.

Bendz, M. (2011). Forestry education and research. In Jansson, U. (ed.). *Agriculture and Forestry in Sweden since 1900: A Cartographic Description*. pp. 178–181. Stockholm: National Atlas of Sweden.

Berglund, E. (2000). Forestry expertise and national narratives: some consequences for old-growth conflicts in Finland. *Worldviews: Global Religions, Culture, and Ecology*, 4(1), 47–67.

Berglund, E. (2001). Facts, beliefs and biases: perspectives on forest conservation in Finland. *Journal of Environmental Planning and Management*, 44(6), 833–849.

Berkes, F. (1999). *Sacred ecology: traditional ecological knowledge and management systems*. Philadelphia and London: Taylor & Francis.

Berkes, F. (2011). Implementing ecosystem-based management: Evolution or revolution? *Fish and Fisheries*, 13(4), 465–476.

Bernes, C. and Lundgren, L. J. (2009). *Use and misuse of nature's resources: An environmental history of Sweden.* Stockholm: Swedish Environmental Protection Agency.

Betänkande med förslag till skogsvårdslag m. m. (1946). Stockholm: Statens offentliga utredningar 1946:41.

Bevir, M. (2013). *Governance: A very short introduction.* Oxford: Oxford University Press.

Biggs, R., Schlüter, M., and Schoon, M. L. (2015). *Principles for Building Resilience: Sustaining Ecosystem Services in Social-Ecological Systems.* Cambridge: Cambridge University Press.

Binder, C. R., Hinkel, J., Bots P. W. G., and Pahl-Wostl, C. (2013). Comparison of frameworks for analyzing social-ecological systems. *Ecology and Society, 18*(4), 26.

Bjärstig, T., Sandström, C., Lindqvist, S. and Kvastegård, E. (2014). Partnerships implementing ecosystem-based moose management in Sweden. *International Journal of Biodiversity Science, Ecosystem Services & Management, 10*(3), 228–239.

Bjärstig, T., and Kvastegård, E. (2016). Forest social values in a Swedish rural context: The private forest owners' perspective. *Forest Policy and Economics, 65,* 17–24.

Bjärstig, T. and Sandström, C. (2017). Public-private partnerships in a Swedish rural context: A policy tool for the authorities to achieve sustainable rural development? *Journal of Rural Studies, 49,* 58–68.

Björklund, J., Agnoletti, M., and Anderson, S. (2000). Exploiting the last phase of the North European Timber Frontier for the international market 1890–1914: an economic-historical approach. In *Forest history: international studies on socio-economic and forest ecosystem change. Report No. 2 of the IUFRO Task Force on Environmental Change,* (pp. 171–184). Florence, Italy: Department of Forestry and Environmental Sciences.

Booth, E. G., Qiu, J., Carpenter, S. R., Schatz, J., Chen, X., Kucharik, C. J., Loheide, S. P., Motew, M. M., Seifert, J. M., and Turner, M. G. (2016). From qualitative to quantitative environmental scenarios: Translating storylines into biophysical modeling inputs at the watershed scale. *Environmental Modelling & Software, 85,* 80–97.

Börjeson, L., Höjer, M., Dreborg, K. H., Ekvall, T., and Finnveden, G. (2006). Scenario types and techniques: towards a user's guide. *Futures, 38*(7), 723–739.

Bormann, F. H., and Likens, G. E. (1979). *Pattern and process in a forested ecosystem: disturbance, development, and the steady state based on the Hubbard Brook ecosystem study.* New York: Springer-Verlag.

Borowy, I. (2015). *Defining sustainable development for our common future.* Abingdon, UK, and New York: Routledge.

Bose, A. K., Harvey, B. D., Brais, S., Beaudet, M., and Leduc, A. (2014). Constraints to partial cutting in the boreal forest of Canada in the context of natural disturbance-based management: a review. *Forestry, 87*(1), 11–28.

Bostedt, G., Widmark, C., Andersson, M., and Sandström, C. (2015). Measuring transaction costs for pastoralists in multiple land use situations: reindeer husbandry in northern Sweden. *Land Economics, 91*(4), 704–722.

Boström, M., Lidskog, R., and Uggla, Y. (2017). A reflexive look at reflexivity in environmental sociology. *Environmental Sociology, 3*(1), 6–16.

Boucher, D., and Kelly, P. J. (1994). *The social contract from Hobbes to Rawls.* London: Routledge.

Bowler, P. J. (1992). *The Fontana history of environmental sciences.* London: Fontana.

British Columbia Forestry Roundtable. (2009). Moving toward a high value, globally competitive, sustainable forest industry. Victoria, British Columbia: Ministry of Forests and Range, www.for.gov.bc.ca/mof/forestry_roundtable/Moving_Toward_a_Globally_Competitive_ Forest_Industry.pdf (accessed 17 March 2017).

Brown, N., Rappert, B., and Webster, A. (eds). (2000). *Contested futures: a sociology of prospective techno-science.* Farnham, UK: Ashgate Publishing.

Brukas, V., and Weber, N. (2009). Forest management after the economic transition – at the crossroads between German and Scandinavian traditions. *Forest Policy and Economics, 11*(8), 586–592.

Brundtland, G.H. and Khalid, M. (1987). *Our common future.* World Commission on Environment and Development. Nairobi: United Nations Environment Programme.

Bruno, K. (2016). *Exporting agrarian expertise: development aid at the Swedish University of Agricultural Sciences and its predecessors, 1950–2009.* Diss. Uppsala: Sveriges lantbruksuniversitet.

Bürgi, M., and Schuler, A. (2003). Driving forces of forest management–an analysis of regeneration practices in the forests of the Swiss Central Plateau during the 19th and 20th century. *Forest Ecology and Management, 176*(1), 173–183.

Burnam-Fink, M. (2015). Creating narrative scenarios: Science fiction prototyping at emerge. *Futures, 70,* 48–55.

Bush, T. (2010). Biodiversity and sectoral responsibility in the development of Swedish forestry policy, 1988–1993. *Scandinavian Journal of History, 35*(4), 471–498.

Carlsson, J., Eriksson, L. O., Öhman, K., and Nordström, E. M. (2015). Combining scientific and stakeholder knowledge in future scenario development–A forest landscape case study in northern Sweden. *Forest Policy and Economics, 61,* 122–134.

Carlsson, J. (2017). *Participatory scenario analysis in forest resource management: exploring methods and governance challenges from a rural landscape perspective.* Diss. Umeå: Sveriges lantbruksuniv.

CBD (Convention on Biological Diversity). (1992). Convention on Biological Diversity. United Nations, *www.cbd.int/doc/legal/cbd-en.pdf* (accessed 21 February 2017).

CBD (Convention on Biological Diversity). (1995). Malawi Principles. Nairobi, Kenya: Convention on Biological Diversity, United Nations Environment Programme, www.cbd.int/ecosystem/principles.shtml

CBD (Convention on Biological Diversity). (2004). Decision VII/11 Ecosystem Approach. Decision adopted by the Conference of Parties to the Convention on Biological Diversity at its seventh meeting, 9–20 and 27 February 2004. Kuala Lumpur, Malaysia: www.cbd.int/doc/decisions/cop-07/cop-07-dec-11-en.doc (accessed 21 February 2017).

CBD (Convention on Biological Diversity). (2012) Ecosystem Approach. Nairobi, Kenya: Convention on Biological Diversity, United Nations Environment Programme, www.cbd.int/ecosystem/

Chazdon, R. L., Brancalion, P. H., Laestadius, L., Bennett-Curry, A., Buckingham, K., Kumar, C., Moll-Rocek, J., Guimaraes Vieria, I. C., and Wilson, S. J. (2016). When is a forest a forest? Forest concepts and definitions in the era of forest and landscape restoration. *Ambio, 45*(5), 538–550.

Christensen, N. L., Bartuska, A. M., Brown, J. H., Carpenter, S., D'Antonio, C., Francis, R., Franklin, J. F., MacMahon, J. A., Noss, R. F., Parsons, D. J., Peterson, C. H., Turner, M. G., and Woodmansee, R. G. (1996). The report of the Ecological Society of America committee on the scientific basis for ecosystem management. *Ecological applications*, 6(3), 665–691.

Cintas, O., Berndes, G., Hansson, J., Poudel, B. C., Bergh, J., Börjesson, P., Lundmark, T., and Nordin, A. (2016). The potential role of forest management in Swedish scenarios towards climate neutrality by mid century. *Forest Ecology and Management, 383*, 73–84.

Cirkulärskrivelse. 1950. Domänverket 1950:1.

Claesson, S., Duvemo, K., Lundström, A., and Wikberg, P-E. (2015). Skogliga konsekvensanalyser 2015 – SKA 2015. Report 1873, Skogsstyrelsen, Jönköping, Sweden.

Clark, W. C., and Dickson, N. M. (2003). Sustainability science: the emerging research program. *Proceedings of the National Academy of Sciences, 100*(14), 8059–8061.

Coleman, D. C. (2010). *Big ecology: The emergence of ecosystem science*. Berkeley: University of California Press.

Corezzola, S., D'Andrea, E., and Zapponi, L. (2016). Indicators of sustainable forest management: a European overview. *Indicators of Sustainable Forest Management: Application and Assessment, 40*(1), 32–35.

Costanza R., D'Arge R., deGroot R., Farber S., Grasso M., Hannon B., Limburg K., Naeem S., Oneill R. V., Paruelo J., Raskin R. G., Sutton P., and van den Belt M. (1997) The value of the world's ecosystem services and natural capital. *Nature 387*, 253–260.

Cowling, R. M., Egoh, B., Knight, A. T., O'farrell, P. J., Reyers, B., Rouget, M., Roux, D.j., Wetz, A., and Wilhelm-Rechman, A. (2008). An operational model for mainstreaming ecosystem services for implementation. *Proceedings of the National Academy of Sciences, 105*(28), 9483–9488.

Dahlberg, A., Thor, G., Allmér, J., Jonsell, M., Jonsson, M., and Ranius, T. (2011). Modelled impact of Norway spruce logging residue extraction on biodiversity in Sweden. *Canadian Journal of Forest Research, 41*(6), 1220–1232.

Daily, G. C. (ed.). (1997). *Nature's services: societal dependence on natural eco-systems*. Washington, DC: Island Press.

Daily G. C. A., S., Ehrlich, P. R., Goulder, L., Lubchenco, J., Matson, P. A., Mooney, H. A., Postel, S., Schneider, S. H., Tilman, D., and Woodwell, G. M. (1997) Ecosystem Services: Benefits Supplied to Human Societies by Natural Ecosystems. *Issues in Ecology, 2* (18).

Dargavel, J., and Johann, E. (2013). *Science and hope: a forest history*. Cambridge: The White Horse Press.

Dargavel, J. (2010). Netting the Global Forest: Attempts at Influence. *Global Environment, 3*(5), 127–158.

Darwin, C. (1859). *On the Origin of Species*. 1st edn. London: John Murray, Albemarle Street.

Dauvergne, P., and Lister, J. (2011). *Timber*. Cambridge: Polity Press.

Daw, T.M., S. Coulthard, W.W.L. Cheung, K. Brown, C. Abunge, D. Galafassi, G.D. Peterson, T.R. McClanahanh, J.O. Omukoto, and L. Munyi. 2015. Evaluating taboo trade-offs in ecosystems services and human well-being. *Proceedings of the National Academy of Sciences (PNAS), 112*(22), 6949–6954.

148 *References*

Decker, D.J., Riley S. J. and Siemer. W.F. (2012) *Human Dimensions of Wildlife Management*. Baltimore, MD: The John Hopkins University Press.
Denhardt, J. V., and Denhardt, R. B. (2007). *The new public service: serving, not steering*. Armonk, NY: M.E. Sharpe.
De Lucia, V. (2015). The ecosystem approach between ecocentrism and anthropocentrism. *SSRN Electronic Journal*. doi:10.2139/ssrn.2520649.
Di Gasper, S. W. (2008). *Natural resource management in an institutional disorder: the development of adaptive co-management systems of moose in Sweden*. Diss. Luleå: Luleå tekniska universitet.
Donner-Amnell, J. (2004). The emergence of two national concepts and their convergence toward a common Nordic regime in the global forest industry. In Lehtinen, A. A., Donner-Amnell, J., and Sæther, B. (eds). *Politics of forests: Northern forest-industrial regimes in the age of globalization*. Burlington, VT: Ashgate Publishing.
Dreborg, K. H. (1996). Essence of backcasting. *Futures*, 28(9), 813–828.
Drössler, L., Nilsson, U., and Lundqvist, L. (2014). Simulated transformation of even-aged Norway spruce stands to multi-layered forests: an experiment to explore the potential of tree size differentiation. *Forestry*, 87(2), 239–248.
Dryzek, J. S. (2000). *Deliberative democracy and beyond: liberals, critics, contestations*. Oxford: Oxford University Press.
Dryzek, J. S., and Pickering, J. (2017). Deliberation as a catalyst for reflexive environmental governance. *Ecological Economics*, 131, 353–360.
Duerr, W. A., and Duerr, J. B. (1975). The role of faith in forest management. In Ramsey, F., and Duerr, W. A. (eds). *Social sciences in forestry: a book of readings* (pp. 30–41). Philadelphia, PA: W.B. Saunders Company.
Eckerberg, K. (1990). *Environmental protection in Swedish forestry*. Aldershot: Avebury.
Eckerberg, K. (1995). Multiple-use forestry administration, legislation and interest groups. In *Multiple-use forestry in the Nordic countries*. Hytönen, M. (ed.), (pp. 357–390). Vantaa, Finland: Finnish Forest Research Institute.
Eckerberg, K., and Joas, M. (2004). Multi-level environmental governance: a concept under stress?. *Local Environment*, 9(5), 405–412.
Eckerberg, K. (2015). Future forest governance: Multiple challenges, diverging responses. In Westholm, E., Beland Lindahl, K., and Kraxner, F. (eds). *The Future Use of Nordic Forests* (pp. 83–97). Heidelberg, New York, Dordrecht, London: Springer International Publishing.
Egan Sjölander, A., Ekerholm, H., Eklöf, J., Lång, H., Mårald, E., Nordlund, C., and Sundin, B. (2014). *Motorspriten kommer! En historia om etanol och andra alternativa drivmedel*. Möklinta: Gidlunds förlag.
Elfving, B., Ericsson, T., and Rosvall, O. (2001). The introduction of lodgepole pine for wood production in Sweden–a review. *Forest Ecology and Management*, 141(1), 15–29.
Eliasson, P. (1997). Från agrart utmarksbruk till industriellt skogsbruk – en långdragen historia. In Östlund, L (ed.). *Människan och skogen* (pp. 46–70; 139–141). Stockholm: Nordiska Museet.
Eliasson, P. (2002). *Skog, makt och människor: En miljöhistoria om svensk skog 1800–1875*. Stockholm: Kungliga Skogs- och Lantbruksakademien.
Eliasson, P. (2011). The state-owned forests: silviculture, mechanization and institutional change. In Antonson, H. and Jansson, U. (eds). *Agriculture and forestry*

in Sweden since 1900: geographical and historical studies (pp. 390–405). Stockholm: Royal Swedish Academy of Agriculture and Forestry.

Elzinga A. (1985). Research, bureaucracy and the drift of epistemic criteria. In Wittlock, B. and Elzinga, A. (eds). *The university research system, the public policies of the home of scientists* (pp. 191–220). Stockholm: Almqvist & Wiksell International.

Emerson, K., Nabatchi, T., and Balogh, S. (2011). An integrative framework for collaborative governance. *Journal of Public Administration Research and Theory*, 22(1), 1–29.

Enander, K-G. (2007). *Skogsbruk på samhällets villkor: skogsskötsel och skogspolitik under 150 år*. Umeå: Institutionen för skogen ekologi och skötsel, Sveriges lantbruksuniversitet.

Enander, K-G. (2011). Forest policy in the 20th century. In Jansson, U. (ed.). *Agriculture and Forestry in Sweden since 1900: A Cartographic Description* (pp. 118–121). Stockholm: National Atlas of Sweden.

Engel, S., Pagiola, S., and Wunder, S. (2008). Designing Payments for Environmental Services in Theory and Practice – An Overview of the Issues. *Ecological Economics* 65, 663–674.

Erefur, C., Bergsten, U., Lundmark, T., and de Chantal, M. (2011). Establishment of planted Norway spruce and Scots pine seedlings: effects of light environment, fertilisation, and orientation and distance with respect to shelter trees. *New Forests*, 41(2), 263–276.

Eriksson, L. (2015). The importance of threat, strategy, and resource appraisals for long-term proactive risk management among forest owners in Sweden. *Journal of Risk Research*, 1–19.

Eriksson, L. (2017). Components and Drivers of Long-term Risk Communication: Exploring the Within-Communicator, Relational, and Content Dimensions in the Swedish Forest Context. *Organization & Environment*, 30(2), 162–179.

Espmark, K. (2017). *Debatten om hyggesfritt skogsbruk i Sverige: En analys av begrepp och argument i svenskt pressmaterial 1994–2013*. Future Forests Rapportserie 2017:2. Sveriges lantbruksuniversitet, Umeå, www.slu.se/futureforests

European Commission. (2012). *Commission staff working document accompanying the document innovating for sustainable growth: a bioeconomy for Europe*. Brussels: European Commission.

European Commission. (2015). The 2030 agenda for sustainable development. http://ec.europa.eu/europeaid/policies/european-development-policy/2030-agenda-sustainable-development (accessed 11 November 2016).

Falkman, L. B. 1852. *Om swenska skogarnas nuwarande tillstånd och deras inflytande på landets framtid*. Stockholm.

FAO (Food and Agriculture Organization of the United Nations) Fisheries Department. (2003). The ecosystem approach to fisheries. *FAO Technical Guidelines for Responsible Fisheries*, 4(2), 112.

FAO (Food and Agriculture Organization of the United Nations). (2010). Global forest resources assessment 2010: main report. Rome: Food and Agriculture Organization of the United Nations.

FAO (Food and Agriculture Organization of the United Nations). (2011). What is an outlook study?, www.fao.org/forestry/outlook/en/

Fareld, V. (2016). (In) between the living and the dead: new perspectives on time in history. *History Compass*, 14(9), 430–440.

Farnham, T. J. (2007). *Saving nature's legacy: origins of the idea of biological diversity.* New Haven, CT: Yale University Press.

Fältbiologerna. (1973). *Skogsbruk och ekologi: Fakta om skogen och skogsbrukets miljöeffekter.* Stockholm: Natur och Kultur.

Felton, A., Lindbladh, M., Brunet, J., and Fritz, Ö. (2010). Replacing coniferous monocultures with mixed-species production stands: an assessment of the potential benefits for forest biodiversity in northern Europe. *Forest Ecology and Management, 260*(6), 939–947.

Folke, C., Carpenter, S., Elmqvist, T., Gunderson, L., Holling, C. S., and Walker, B. (2002). Resilience and sustainable development: building adaptive capacity in a world of transformations. *Ambio, 31*(5), 437–440.

Folke, C. (2004). Traditional knowledge in social–ecological systems. *Ecology and Society, 9*(3), www.ecologyandsociety.org/vol9/iss3/art7/

Folke, C. (2006). Resilience: the emergence of a perspective for social–ecological systems analyses. *Global Environmental Change, 16*(3), 253–267.

Folke, C., Hahn, T., Olsson P., and Norberg, J. (2010). Adaptive governance of social-ecological systems. *Annual Review of Environment and Resources, 30*, 441–473.

Folke, C., Biggs, R., Norström, A., Reyers, B., and Rockström, J. (2016). Social-ecological resilience and biosphere-based sustainability science. *Ecology and Society, 21*(3).

Ford, C. (2016). *Natural interests: the contest over environment in modern France.* Cambridge, MA: Harvard University Press.

Forest Europe. (2011). *State of Europe's Forests 2011.* Ministerial Conference on the Protection of Forests in Europe, Forest Europe Liaison Unit Oslo.

Forest Europe. (2015). *State of Europe's Forests 2015.* Ministerial Conference on the Protection of Forests in Europe, Forest Europe Liaison Unit Madrid.

Formas. (2012). *Swedish research and innovation strategy for a bio-based economy*; R3:2012. Stockholm: Swedish Research Council for Environment, Agricultural Sciences and Spatial Planning (FORMAS).

Freeman, R. (2002). The ecofactory: The United States Forest Service and the political construction of ecosystem management. *Environmental history, 7*(4), 632–658.

Fridman, J., Holm, S., Nilsson, M., Nilsson, P., Ringvall, A., and Ståhl, G. (2014). Adapting National Forest Inventories to changing requirements – the case of the Swedish National Forest Inventory at the turn of the 20th century. *Silva Fennica, 48*(3). doi:10.14214/sf.1095

Future Forests. (2013). Programme plan 2013–2016. www.slu.se/globalassets/ew/org/centrb/f-for/pdf/future-forests-programplan-2013–2016.pdf

Gamfeldt, L., Snäll, T., Bagchi, R., Jonsson, M., Gustafsson, L., Kjellander, P., Ruiz-Jaen, M. C., Fröberg, M., Stendahl, J., Philipson, C. D., Andersson, E., Westerlund, B., Andrén, H., Moberg, F., Moen, J., and Mikusiński, G. (2013). Higher levels of multiple ecosystem services are found in forests with more tree species. *Nature Communications, 4*, 1340.

Gardner, R. (2009). Constructing a technological forest: nature, culture, and tree-planting in the Nebraska sand hills. *Environmental History, 14*(2), 275–297.

Geels, F. W. (2002). Technological transitions as evolutionary reconfiguration processes: a multi-level perspective and a case-study. *Research policy, 31*(8), 1257–1274.

Geels, F.W. (2005). *Technological Transitions and System Innovations: A Co - Evolutionairy and Socio-Technical Analysis*. Cheltenham, UK: Edward Elgar.

Giddens, A. (1994). Living in a post-traditional society. In Beck, U., Giddens, A., and Lash, S. (eds). *Reflexive modernization: politics, tradition and aesthetics in the modern social order*. Cambridge: Polity Press.

Giddens, A. (1999). Risk and responsibility. *The Modern Law Review*, 62(1), 1–10.

Glasbergen, P. (2007). Setting the scene: the partnership paradigm in the making. *Partnerships, governance and sustainable development: reflections on theory and practice*, 1–25.

Glasbergen, P. (ed.). (2012). *Managing environmental disputes: network management as an alternative*, 5. Springer Science & Business Media

Govt. Prop. (2013). En svensk strategi för biologisk mångfald och ekosystemtjänster, p. 141, www.riksdagen.se/sv/Dokument-Lagar/Forslag/Propositioner-ochskrivelser/En-svensk-strategi-for-biologi_H103141/?text=true (accessed 17 March 2017).

Griggs, D., Stafford-Smith, M., Gaffney, O., Rockström, J., Öhman, M. C., Shyamsundar, P., Steffen, W., Glaser, G., Norichika, K., and Noble, I. (2013). Policy: sustainable development goals for people and planet. *Nature*, 495(7441), 305–307.

Grumbine, R. E. (1994). What is ecosystem management? *Conservation Biology*, 8(1), 27–38.

Grumbine, R. E. (1997). Reflections on 'What is Ecosystem Management?. *Conservation Biology*, 11(1), 41–47.

Guivarch, C., Rozenberg, J., and Schweizer, V. (2016). The diversity of socio-economic pathways and CO 2 emissions scenarios: Insights from the investigation of a scenarios database. *Environmental Modelling & Software*, 80, 336–353.

Gunderson, L. H. (2000). Ecological resilience–in theory and application. *Annual review of ecology and systematics*, 31(1), 425–439.

Gustafsson, L., Kouki, J., and Sverdrup-Thygeson, A. (2010). Tree retention as a conservation measure in clear-cut forests of northern Europe: a review of ecological consequences. *Scandinavian Journal of Forest Research*, 25(4), 295–308.

Gustafsson, L., Baker, S. C., Bauhus, J., Beese, W. J., Brodie, A., Kouki, J., . . . and Neyland, M. (2012). Retention forestry to maintain multifunctional forests: a world perspective. *BioScience*, 62(7), 633–645.

Gustafsson, K. M., and Lidskog, R. (2013). Boundary work, hybrid practices, and portable representations: an analysis of global and national coproductions of red lists. *Nature and Culture*, 8(1), 30–52.

Hahn, T. (2000). *Property rights, ethics, and conflict resolution: Foundations of the Sami economy in Sweden*. Diss. Uppsala: Sveriges lantbruksuniversitet.

Hajer, M. A. (1995). *The politics of environmental discourse: Ecological modernization and the policy process*. Oxford: Clarendon Press.

Hall, M. (2005). *Earth repair: A transatlantic history of environmental restoration*. Charlottesville: University of Virginia Press.

Hartog, F. (2015). *Regimes of historicity: Presentism and experiences of time*. New York, and Chichester, UK: Columbia University Press.

Harvey, D. (2007). *A brief history of neoliberalism*. New York: Oxford University Press.

Harwood, J. (2005). *Technology's dilemma: Agricultural colleges between science and practice in Germany, 1860–1934*. Oxford: P. Lang.

Harwood, J. (2010). Understanding academic drift: On the institutional dynamics of higher technical and professional education. *Minerva*, 48(4), 413–427.

Hayter, R., and Soyez, D. (1996). Clearcut issues: German environmental pressure and the British Columbia forest sector. *Geographische Zeitschrift*, 143–156.

HELCOM. (2008). Guidelines for HELCOM coastal fish monitoring sampling methods. www.helcom.fi/helcom-at-work/events/helcom-stakeholder-conferences-until-2010/2008-conference

Hellström, E., and Rytilä, T. (1998). *Environmental forest conflicts in France and Sweden: Struggling between local and international pressure.* Joensuu, Finland: European Forest Institute.

Hicks, C. C., Levine, A., Agrawal, A., Basurto, X., Breslow, S. J., Carothers, C., Charnley, S., Coulthard, S., Dolsak, N., Noatuto, J., Garcia-Quijano, C., Mascia, M. B., Norman, K., Poe, M. R., Satterfield, T., St. Martin, K., and Levin, P. S. (2016). Engage key social concepts for sustainability. *Science*, 352(6281), 38–40.

Higgs, E., Falk, D. A., Guerrini, A., Hall, M., Harris, J., Hobbs, R. J., Jackson, S. T., Rhemtulla, J. M., and Throop, W. (2014). The changing role of history in restoration ecology. *Frontiers in Ecology and the Environment*, 12(9), 499–506.

Hilson, M. (2008). *The Nordic model: Scandinavia since 1945.* London: Reaktion Books.

Hirt, P. W. (1994). *A conspiracy of optimism: Management of the national forests since World War Two.* Lincoln: University of Nebraska Press.

Höjer, M., and Mattsson, L. G. (2000). Determinism and backcasting in future studies. *Futures*, 32(7), 613–634.

Holling, C. S. (1973). Resilience and stability of ecological systems. *Annual Review of Ecology and Systematics*, 4(1), 1–23.

Holling, C. S. (1978). *Adaptive environmental assessment and management.* Chichester, UK: John Wiley & Sons.

Holling, C. S., and Meffe, G. K. (1996). Command and control and the pathology of natural resource management. *Conservation Biology*, 10(2), 328–337.

Holmgren, L., Sandström, C., and Zachrisson, A. (2016). Protected area governance in Sweden: New modes of governance or business as usual? *Local Environment* 22(1), 22–37.

Holmström, E., Hjelm, K., Karlsson, M., and Nilsson, U. (2016). Scenario analysis of planting density and pre-commercial thinning: will the mixed forest have a chance? *European Journal of Forest Research*, 135(5), 885–895.

Hölzl, R. (2010). Historicizing sustainability: German scientific forestry in the eighteenth and nineteenth centuries. *Science as Culture*, 19(4), 431–460.

Hongslo, E., Hovik, S., Zachrisson, A., and Aasen Lundberg, A. K. (2016). Decentralization of conservation management in Norway and Sweden–Different translations of an international trend. *Society & Natural Resources*, 29(8), 998–1014.

Hood, C. (1995). The 'New Public Management' in the 1980s: variations on a theme. *Accounting, organizations and society*, 20(2), 93–109.

Hooghe, L., and Marks, G. (2001). *Multi-level governance and European integration.* Oxford: Rowman & Littlefield.

Hoogstra-Klein, M. A., Hengeveld, G. M., and de Jong, R. (2016). Analysing scenario approaches for forest management–one decade of experiences in Europe. *Forest Policy and Economics*. http://dx.doi.org/10.1016/j.forpol.2016.10.002

Horstkotte, T., Sandström, C., and Moen, J. (2014). Exploring the multiple uses of boreal landscapes in northern Sweden: The importance of social-ecological diversity for mobility and flexibility. *Human Ecology*, 42(5), 671–682.

Hovik, S., Harvold, K., and Joas, M. (2009). New approaches to managing protected areas in the Nordic countries. *Local Environment*, *14*(3), 215–220, DOI: 10.1080/13549830802692781.

Hovik, S., Sandström, C., and Zachrisson, A. (2010). Management of protected areas in Norway and Sweden: challenges in combining central governance and local participation. *Journal of Environmental Policy & Planning*, *12*(2), 159–177.

Howlett, M. and Rainer, J. (2006). Globalization and Governance Capacity: Explaining Divergence in National Forest Programmes as Instances of 'Next-Generation' Regulation in Canada and Europe. *Governance*, *19*(2), 251–275.

Howlett, M., Mukherjee, I., and Rayner, J. (2014). The elements of effective program design: a two-level analysis. *Politics and Governance*, *2*(2), 1–12.

Hughes, T., Bellwood, D., Folke, C., Steneck, R., and Wilson, J. (2005). New paradigms for supporting the resilience of marine ecosystems. *Trends in Ecology & Evolution*, *20*(7), 380–386.

Humphreys, D. (2009). Science, knowledge, values and forest policy. *Journal of Integrative Environmental Sciences*, *6*(3), 157–161.

Humphreys, D. (2014). *Forest politics: the evolution of international cooperation.* Abingdon, UK, and New York: Routledge.

Hunt, L. (2002). Against presentism. *Perspectives*, *40*(5), 7–9.

Hurmekoski, E., and Hetemäki, L. (2013). Studying the future of the forest sector: review and implications for long-term outlook studies. *Forest Policy and Economics*, *34*, 17–29.

Hytönen, M. (ed.). (1995). *Multiple-use forestry in the Nordic countries.* Vantaa, Finland: METLA, The Finnish Forest Research Institute.

IPCC. (2013). Climate change 2013: the physical science basis. Contribution of working group I to the fifth assessment report of the intergovernmental panel on climate change (Stocker, T.F., Qin, D., Plattner, G. K., Tignor, M., Allen, S. K., Boschung, J., Nauels, A., Xia, Y., Bex, V., and Midgley, P. M. (eds). Cambridge and New York: Cambridge University Press.

Ison, R., and Watson, D. (2007). Illuminating the possibilities for social learning in the management of Scotland's water. *Ecology and Society*, *12*(1), Art-21.

Jasanoff, S. (2003). Technologies of humility: citizen participation in governing science. *Minerva*, *41*(3), 223–244.

Johansson, J. (2012). Challenges to the legitimacy of private forest governance–the development of forest certification in Sweden. *Environmental Policy and Governance*, *22*(6), 424–436.

Johansson, J. (2013). *Constructing and contesting the legitimacy of private forest governance – the case of forest certification in Sweden.* Research report 2013:1. Umeå: Umeå University, Department of Political Science.

Johansson, J. (2014). Why do forest companies change their CSR strategies? Responses to market demands and public regulation through dual-certification. *Journal of Environmental Planning and Management*, *57*(3), 349–368.

Johansson, J., and Keskitalo, E. C. H. (2014). Coordinating and implementing multiple systems for forest management: implications of the regulatory framework for sustainable forestry in Sweden. *Journal of Natural Resources Policy Research*, *6*(2–3), 117–133.

Johansson, J. (2016). Participation and deliberation in Swedish forest governance: The process of initiating a National Forest Program. *Forest Policy and Economics*, *70*, 137–146.

Johansson, T., Hjältén, J., Olsson, J., Dynesius, M., and Roberge, J. M. (2016). Long-term effects of clear-cutting on epigaeic beetle assemblages in boreal forests. *Forest Ecology and Management*, *359*, 65–73.

Jonsson, R. (2011). Trends and possible future developments in global forest-product markets–Implications for the Swedish forest sector. *Forests*, *2*(1), 147–167.

Jonsson, R. (2013). How to cope with changing demand conditions – The Swedish forest sector as a case study: an analysis of major drivers of change in the use of wood resources. *Canadian Journal of Forest Research*, *43*(999), 405–418.

Jörnmark, J. (2004). *Skogen, staten och kapitalisterna. Skapande förstörelse i svensk basindustri 1810–1950*. Lund: Studentlitteratur.

Josefsson, T. (2009). *Pristine forest landscapes as ecological references: human land use and ecosystem change in boreal Fennoscandia*. Diss. Umeå: Sveriges lantbruksuniversitet.

Kaijser, A., and Tiberg, J. (2000). From operations research to futures studies: the establishment, diffusion and transformation of the systems approach in Sweden 1945–1980. In Hughes, A.C., and Parke Hughes, T. (eds). *Systems, Experts and Computers: The Systems Approach in Management and Engineering World War II and After*, (pp. 385–412). Cambridge, MA: MIT Press.

Kaiserfeld, T. (2013). Why new hybrid organizations are formed: Historical perspectives on epistemic and academic drift. *Minerva*, *51*(2), 171–194.

Kardell, L. (2004). *Svenskarna och skogen. D. 2, Från baggböleri till naturvård*. Jönköping: Skogsstyrelsens förlag.

Kardell, Ö. (2016). Swedish Forestry, Forest Pasture Grazing by Livestock, and Game Browsing Pressure Since 1900. *Environment and History*, *22*(4), 561–587.

Keck, M. E., and Sikkink, K. (2014). *Activists beyond borders: Advocacy networks in international politics*. Ithaca, NY: Cornell University Press.

Kemp, R. (1994). Technology and the transition to environmental sustainability: the problem of technological regime shifts. *Futures*, *26*(10), 1023–1046.

Kemp, R., and Loorbach, D. (2006). Transition management: a reflexive governance approach. *Reflexive Governance for Sustainable Development*, Cheltenham, UK, and Northampton, MA: Edward Elgar.

Kemp, R., Loorbach, D. and Rotmans, J. (2007a). Transition management as a model for managing processes of co-evolution towards sustainable development. *International Journal of Sustainable Development and World Ecology*, *14*(1), 78–91.

Kemp, R., Loorbach, D. and Rotmans, J. (2007b). Assessing the Dutch energy transition policy: how does it deal with dilemmas of managing transitions? *Journal of Environmental Policy and Planning*, *9*(3/4), 315–331.

Kenny, M., and Meadowcroft, J. (1999). *Planning sustainability: the implications of sustainability for public planning policy*. London: Routledge.

Keskitalo, E. C. H., Bergh, J., Felton, A., Björkman, C., Berlin, M., Axelsson, P., Ring, E., Ågren, A., Roberge, J-M., Klapwijk, M. J. and Boberg, J. (2016). Adaptation to Climate Change in Swedish Forestry. *Forests*, *7*(2), 28.

Keskitalo, E. C. H., Horstkotte, T., Kivinen, S., Forbes, B., and Käyhkö, J. (2016). 'Generality of mis-fit'? The real-life difficulty of matching scales in an inter-connected world. *Ambio*, *45*(6), 742–752.

Kjær, A. M. (2004). *Governance*. Malden, MA: Polity Press.

Klang, F., and Ekö, P. M. (1999). Tree properties and yield of Picea abies planted in shelterwoods. *Scandinavian Journal of Forest Research*, *14*(3), 262–269.

Kleinschmit, D., Lindstad, B. H., Thorsen, B. J., Toppinen, A., Roos, A., and Baardsen, S. (2014). Shades of green: A social scientific view on bioeconomy in the forest sector. *Scandinavian Journal of Forest Research*, 29(4), 402–410.

Klinke, A. (2009). Reflexive Governance Coping with Complexity, Uncertainty and Ambiguity: Paper for the XXI IPSA World Congress of Political Science. Santiago.

Korosuo, A., Sandström, P., Öhman, K., and Eriksson, L. O. (2014). Impacts of different forest management scenarios on forestry and reindeer husbandry. *Scandinavian Journal of Forest Research*, 29(sup1), 234–251.

Koselleck, R. (2004). *Futures past: on the semantics of historical time.* New York, and Chichester, UK: Columbia University Press.

Kotilainen, J., and Rytteri, T. (2011). Transformation of forest policy regimes in Finland since the 19th century. *Journal of Historical Geography*, 37(4), 429–439.

Kraxner, F., Nordström, E. M., Havlík, P., Gusti, M., Mosnier, A., Frank, S., Valin, H., Fritz, S., Fuss, S., Kindermann, G., McCallum, I., Khabarov, N., Böttcher, H., See, L., Kentaro, A., Schmid, E., Obersteiner, M., and Máthég, L. (2013). Global bioenergy scenarios–Future forest development, land-use implications, and trade-offs. *Biomass and Bioenergy*, 57, 86–96.

Kraxner, F., and Nordström, E. M. (2015). Bioenergy futures: A global outlook on the implications of land use for forest-based feedstock production. In. Westholm, E., Lindahl, K. B., and Kraxner, F. (eds). (2015). *The Future Use of Nordic Forests: a global perspective.* Heidelberg, New York, Dordrecht, London: Springer International Publishing.

KSLA (Royal Swedish Academy of Agriculture and Forestry). (2009). *The Swedish Forestry Model.* Stockholm.

Kuglerová, L., Ågren, A., Jansson, R., and Laudon, H. (2014). Towards optimizing riparian buffer zones: Ecological and biogeochemical implications for forest management. *Forest Ecology and Management*, 334, 74–84.

Kungl. Skogs- och lantbruksakademien (2011). *National atlas of Sweden. Agriculture and forestry in Sweden since 1900 a cartographic description* (1st edn). Stockholm: Norstedt.

Kuosa, T. (2011). Evolution of futures studies. *Futures*, 43(3), 327–336.

Kuuluvainen, T., and Ylläsjärvi, I. (2011). On the natural regeneration of dry heath forests in Finnish Lapland: a review of VT Aaltonen (1919). *Scandinavian journal of forest research*, 26(S10), 34–44.

Kuuluvainen, T., Tahvonen, O., and Aakala, T. (2012). Even-aged and uneven-aged forest management in boreal Fennoscandia: a review. *Ambio*, 41(7), 720–737.

Lafferty, W. M. (2004). From environmental protection to sustainable development: the challenge of decoupling through sectoral integration. *Governance for sustainable development: The challenge of adapting form to function*, 191–220.

Lafferty, W. M., and Eckerberg, K. (2013). *From the Earth Summit to Local Agenda 21: working towards sustainable development* (Vol. 12). Abingdoon, UK: Routledge.

Lakoff, A. (2007). Preparing for the next emergency. *Public Culture*, 19(2), 247.

Langhelle, O. (2000). Sustainable development and social justice: expanding the Rawlsian framework of global justice. *Environmental Values*, 9(3), 295–323.

Langston, N. (1995). *Forest dreams, forest nightmares: The paradox of old growth in the Inland West.* Seattle: University of Washington Press.

Larsson, S., Lundmark, T., Mårald, E., and Sandström, C. (2014). *Rapport om det Nationella skogsprogrammets dialogprocess. Future Forests Rapport 2014:2.* Umeå: Swedish University of Agricultural Sciences.

Latour, B. (1987). *Science in action: How to follow scientists and engineers through society*. Cambridge, MA: Harvard University Press.

Latta, G. S., Sjølie, H. K., and Solberg, B. (2013). A review of recent developments and applications of partial equilibrium models of the forest sector. *Journal of Forest Economics, 19*(4), 350–360.

Lauri, P., Forsell, N., Korosuo, A., Havlík, P., and Nordin, A. (2017). Impact of the 2°C target on the global woody biomass use. Submitted manuscript.

Lavsund, S., and Sandegren, F. (1989). Swedish moose management and harvest during the period 1964–1989. *Alces, 25*, 58–62.

Lavsund, S. (2003). *Skogsskötsel och älgskador i tallungskog*. Uppsala: SkogForsk.

Leach, M., Scoones, I., and Stirling, A. (2010). *Dynamic sustainabilities: Technology, environment, social justice*. London: Earthscan.

Lehtinen, A. A., Donner-Amnell, J., and Sæther, B. (2004). *Politics of forests: Northern forest-industrial regimes in the age of globalization*. Burlington, VT: Ashgate Pub.

Lele, S., Springate-Baginski, O., Lakerveld, R., Deb, D., and Dash, P. (2013). Ecosystem Services: Origins, Contributions, Pitfalls, and Alternatives. *Conservation and Society, 11*(4), 343–358.

Lidskog, R. and Sundqvist, G. (2011). Transboundary air pollution policy in transition. In Lidskog, R., and Sundqvist, G. (eds). *Governing the Air: The Dynamics of Science, Policy, and Citizen Interaction*. Cambridge, MA: MIT Press.

Lidskog, R. (2014). Representing and regulating nature: boundary organizations, portable representations, and the science–policy interface. *Environmental Politics, 23*(4), 670–687.

Lidskog, R., and Löfmarck, E. (2015). Managing uncertainty: forest professionals' claim and epistemic authority in the face of societal and climate change. *Risk Management, 17*(3), 145–164.

Lidskog, R., and Löfmarck, E. (2016). Fostering a flexible forest: challenges and strategies in the advisory practice of a deregulated forest management system. *Forest Policy and Economics, 62*, 177–183.

Liesbet, H., and Gary, M. (2003). Unraveling the central state, but how? Types of multi-level governance. *American Political Science Review, 97*(02), 233–243.

Lindahl, K. B. (2008). *Frame analysis, place perceptions and the politics of natural resource management: Exploring a forest policy controversy in Sweden*. Diss. Uppsala: Deptartment of Urban and Rural Development, Swedish University of Agricultural Sciences.

Lindahl, K., and Westholm, E. (2011). Food, paper, wood, or energy? Global trends and future Swedish forest use. *Forests, 2*(1), 51–65.

Lindahl, K. B., and Westholm, E. (2012). Future forests: Perceptions and strategies of key actors. *Scandinavian Journal of Forest Research, 27*(2), 154–163.

Lindahl, K. B. (2015). Actors' perceptions and strategies: Forests and pathways to sustainability. In *The Future Use of Nordic Forests* (pp. 111–124). Heidelberg, New York, Dordrecht, London: Springer International Publishing.

Lindahl, K. B., Sténs, A., Sandström, C., Johansson, J., Lidskog, R., Ranius, T., and Roberge, J-M. (2015). The Swedish forestry model: More of everything? *Forest Policy and Economics*. doi:10.1016/j.forpol.2015.10.012

Lindahl, K. B., Sandström, C., and Sténs, A. (2017). Alternative pathways to sustainability? Comparing forest governance models. *Forest Policy and Economics, 77*, 69–78.

Lindblom, C. E. (1959). The science of 'muddling through'. *Public Administration Review*, 79–88.

Lindenmayer, D., Franklin, J., Lõhmus, A., Baker, S., Bauhus, J., Beese, W., Brodie, A., Kiehl, B., Kouki, J., Martínez Pastur, G., Messier, C., Neyland, M., Palik, B., Sverdup-Theygeson, A., Volney, J., Wayne, A., and Gustafsson, L. (2012). A major shift to the retention approach for forestry can help resolve some global forest sustainability issues. *Conservation Letters*, 5(6), 421–431.

Linder, P. and Östlund, L. (1992). *Förändringar i Sveriges boreala skogar 1870–1991.* Umeå: Swedish University of Agricultural Sciences.

Linder, P., and Östlund, L. (1998). Structural changes in three mid-boreal Swedish forest landscapes, 1885–1996. *Biological Conservation*, 85(1–2), 9–19.

Lindkvist, A., Kardell, Ö., and Nordlund, C. (2011). Intensive forestry as progress or decay? An analysis of the debate about forest fertilization in Sweden, 1960–2010. *Forests*, 2(1), 112–146.

Lindqvist, S., Sandström, C., Bjärstig, T., and Kvastegård, E. (2014). The changing role of hunting in Sweden – from subsistence to ecosystem stewardship? *ALCES* 50, 53–66.

Lindstad, B. H., and Solberg, B. (2012). Influences of international forest policy processes on national forest policies in Finland, Norway and Sweden. *Scandinavian Journal of Forest Research*, 27(2), 210–220.

Linser, S., Wolfslehner, B., and Pülzl. H. (2015). The genesis of the pan-European criteria and indicators and their further development towards emerging policy need. XIV World Forestry Congress, Durban, South Africa, 7–11 September 2015.

Lisberg Jensen, E. (2002). *Som man ropar i skogen: Modernitet, makt och mångfald i kampen om Njakafjäll och i den svenska skogsbruksdebatten 1970–2000.* Diss. Lund: Lund University.

Lisberg Jensen, E. (2006). Sätt stopp för sprutet! Från arbetsmiljöproblem till ekologisk risk i 1970-talets debatt om hormoslyr och DDT i skogsbruket. In Björk, F., Eliasson, P., and Fritzbøger, B. (eds). *Miljöhistoria över gränser* (pp. 197–230). Malmö: Malmö högskola.

Lisberg Jensen, E. (2011). Modern clear-felling: from success story to negotiated solution. In Antonson, H. and Jansson, U. (eds). *Agriculture and Forestry in Sweden since 1900: Geographical and Historical Studies* (pp. 423–441). Stockholm: The Royal Swedish Academy of Agriculture and Forestry.

Liu, J., Li, S., Ouyan, Z, Tam, C., and Chen, X. (2008) Ecological and socioeconomic effects of China's policies for ecosystem services. *Proceedings of the National Academy of Sciences 105*, 9477–9482.

Lockwood, M. (2010). Good governance for terrestrial protected areas: A framework, principles and performance outcomes. *Journal of Environmental Management*, 91(3), 754–766.

Long, R. D., Charles, A., and Stephenson, R. L. (2015). Key principles of marine ecosystem-based management. *Marine Policy*, 57, 53–60.

Lönnroth, M., Johansson, T.B., and Steen, P. (eds). (1980). *Solar versus nuclear: choosing energy futures.* Oxford: Pergamon Press.

Loorbach, D., and Rotmans, J. (2006). Managing transitions for sustainable development. In *Understanding industrial transformation: views from different disciplines* (pp. 187–206). Dordrecht, the Netherlands: Springer.

Loorbach, D. (2007). *Transition management: new mode of governance for sustainable development.* The Netherlands: International Books.

Loorbach, D. (2010). Transition management for sustainable development: a pre-scriptive, complexity-based governance framework. *Governance 23*(1),161–183.

Lorimer, J. (2015). *Wildlife in the Anthropocene*.Minneapolis, MN, and London: University of Minnesota Press.

Lowenthal, D. (2000). *George Perkins Marsh, prophet of conservation*. Seattle: University of Washington Press.

Lowood, H. (1990). The calculating forester: quantification, cameral science, and the emergence of scientific forest management in Germany. In Frängsmyer, T., Heilbron, J.L. and Rider, R.E. (eds). *The Quantifying Spirit in the 18th Century*, pp. 315–342. Berkeley and Los Angeles: University of California.

Lundgren, L. J. (2009). *Staten och naturen: Naturskyddspolitik i Sverige 1869–1935*. Brottby: Kassandra.

Lundmark, H., Josefsson, T., and Östlund, L. (2013). The history of clear-cutting in northern Sweden–driving forces and myths in boreal silviculture. *Forest Ecology and Management, 307*, 112–122.

Lundmark, L., and Rumar, L. (2008). *Mark och rätt i Sameland*. Stockholm: Institutet för rättshistorisk forskning.

Lundmark, L. (2008). *Stulet land: svensk makt på samisk mark*. Stockholm: Ordfront förlag.

Lundmark, T., Bergh, J., Hofer, P., Lundström, A., Nordin, A., Poudel, B. C., Sathre, R., Taverna, R., and Werner, F. (2014). Potential roles of Swedish forestry in the context of climate change mitigation. *Forests, 5*(4), 557–578.

Lundmark, T., Bergh, J., Nordin, A., Fahlvik, N., and Poudel, B. C. (2016). Comparison of carbon balances between continuous-cover and clear-cut forestry in Sweden. *Ambio, 45*(2), 203–213.

Mackinnon, A. (2008). Ecosystem-based management in the central and north coast areas of British Columbia. In McAfee, B., and Malouin C. (eds). *Implementing ecosystem-based management approaches in Canada's forests: a science-policy dialogue*. Ottawa: Natural Resources Canada.

Malmberg, B. (2015). Future forest trends: can we build on demographically based forecasts?. In Westholm *et al.* (eds). *The Future Use of Nordic Forests* (pp. 25–42). Heidelberg, New York, Dordrecht, London: Springer International Publishing.

Mann, M. E. (2013). *The hockey stick and the climate wars: Dispatches from the front lines*. New York, and Chichester, UK: Columbia University Press.

Mårald, E. (2002). Everything circulates: agricultural chemistry and recycling theories in the second half of the nineteenth century. *Environment and History, 8*(1), 65–84.

Mårald, E. (2011). Knowledge in the service of agriculture: knowledge on the borderline between academe and farming. In Antonson, H. and Jansson, U. (eds). *Agriculture and forestry in Sweden since 1900: geographical and historical studies*, pp. 93–108. Stockholm: Royal Swedish Academy of Agriculture and Forestry.

Mårald, E., Sandström, C., Rist, L., Rosvall, O., Samuelsson, L., and Idenfors, A. (2015). Exploring the use of a dialogue process to tackle a complex and contro-versial issue in forest management. *Scandinavian Journal of Forest Research, 30*(8), 749–756.

Mårald, E., and Westholm, E. (2016). Changing approaches to the future in Swedish forestry, 1850–2010. *Nature and Culture, 11*(1), 1–21.

Mårald, E., Langston, N., Sténs, A., and Moen, J. (2016). Changing ideas in forestry: a comparison of concepts in Swedish and American forestry journals during the early twentieth and twenty-first centuries. *Ambio, 45*(2), 74–86.

Marien, M. (2010). Futures-thinking and identity: Why 'Futures Studies' is not a field, discipline, or discourse: a response to Ziauddin Sardar's 'the namesake'. *Futures, 42*(3), 190–194.

Marris, E. (2011). *Rambunctious garden: saving nature in a post-wild world.* New York: Bloomsbury.

Marsh, G. P. (1864/1967). *Man and nature.* Cambridge: Harvard University Press.

Masini, E. (1993). *Why futures studies?* London: Grey Seal Books.

Mather, A. S. (2001). Forests of consumption: postproductivism, postmaterialism, and the postindustrial forest. *Environment and Planning C: Government and policy, 19*(2), 249–268.

McAfee, B. J., and de Camino, R. (2010). Managing Forested Landscapes for Socio-Ecological Resilience. In Mery, G., P. Katila, G. Galloway, R. I. Alfaro, M. Kanninen, M. Lobovikov, and J.Varjo. (eds). *Forests and Society – Responding to Global Drivers of Change: IUFRO World Series Volume 25* (pp.401–440). Vienna: IUFRO.

McGinnis, M. D., and Ostrom, E. (2014). Social-ecological system framework: initial changes and continuing challenges. *Ecology and Society, 19*(2), 30.

McGinnis, M. D. (2016). *Polycentric governance in theory and practice: dimensions of aspirations and practical limitations.* Bloomington, IN: Indiana University.

McNeill, J. R. (2000). *Something new under the sun: an environmental history of the twentieth-century world.* New York: W.W. Norton.

MCPFE. (2003.) Improved pan-European indicators for sustainable forest management. Fourth Ministerial Conference on the Protection of Forests in Europe. Ministerial Conference on the Protection of Forests in Europe, Vienna Liaison Unit, Austria.

MCPFE. (2007). State of Europe's forests 2007. The MCPFE Report on Sustainable Forest Management in Europe. Ministerial Conference on the Protection of Forests in Europe, Liaison Unit Warsaw, Poland.

Meffe, G.K., Nielsen, L.A., Knight, R.L., and Schenborn, D.A. (2002). *Ecosystem Management: Adaptive, Community-Based Conservation.* Washington, DC: Island Press.

MEA (Millennium Ecosystem Assessment). (2005). Ecosystems and human well-being. Washington, DC: Island Press.

Meadowcroft, J. (2007). Who is in charge here? Governance for sustainable development in a complex world. *Journal of Environmental Policy & Planning, 9*(3–4), 299–314.

Meadowcroft, J. (2009) What about the politics? Sustainable development, transition management, and long-term energy transitions. *Policy Science 42,* 323.

Meadows, D. H. (ed.). (1972). *The limits to growth: a report to the club of Rome's project of the predicament of mankind.* New York: New American Library.

Messier, C., Puettmann, K. J., and Coates, K. D. (eds). (2013). *Managing forests as complex adaptive systems: building resilience to the challenge of global change.* Abingdon, UK: Routledge.

Messner, D. (2015). A social contract for low carbon and sustainable development: Reflections on non-linear dynamics of social realignments and technological innovations in transformation processes. *Technological Forecasting and Social Change, 98,* 260–270.

Milestad, R., Svenfelt, Å., and Dreborg, K. H. (2014). Developing integrated explorative and normative scenarios: the case of future land use in a climate-neutral Sweden. *Futures, 60,* 59–71.

Moen, J., Esselin, A., and Nordin, A. (2010). Possible futures, future possibilities. Future Forests Report, www.slu.se/globalassets/ew/org/centrb/f-for/futureforests/ff-scenarier_a4_webb.pdf

Moen, J., Rist, L., Bishop, K., Chapin, F. S., Ellison, D., Kuuluvainen, T., Petersson, H., Puettmann, K. J., Rayner, J., Warkentin, I. G. and Bradshaw, C. J. (2014). Eye on the taiga: removing global policy impediments to safeguard the boreal forest. *Conservation Letters*, 7(4), 408–418.

Morell, M. (2011). Farmland: ownership or leasehold, inheritance or purchase. In Antonsson, H. and Jansson, U. (eds). *Agriculture and Forestry in Sweden since 1900: Geographical and Historical Studies*, pp. 56–73. Stockholm: The Royal Swedish Academy of Agriculture and Forestry.

Mostert, E., Pahl-Wostl, C., Rees, Y., Searle, B., Tàbara, D., and Tippett, J. (2007). Social learning in European river-basin management: barriers and fostering mechanisms from 10 river basins. *Ecology and Society*, 12(1).

Muldoon, R. (2016). *Social contract theory for a diverse world: Beyond tolerance.* New York, and Abingdon, UK: Routledge.

Muro, M., and Jeffrey, P. (2008). A critical review of the theory and application of social learning in participatory natural resource management processes. *Journal of Environmental Planning and Management*, 51(3), 325–344.

Naidoo, R., Balmford, A., Costanza, R., Fisher, B., Green, R. E., Lehner, B., Malcolm, T. R., and Ricketts, T. H. (2008). Global mapping of ecosystem services and conservation priorities. *Proceedings of the National Academy of Sciences*, 105(28), 9495–9500.

Nakicenovic, N., and Swart, R. (2000). *Special Report on Emissions Scenarios.* Cambridge: Cambridge University Press.

Naumann, S., Anzaldua, G., Berry, P., Burch, S., Davis, M., Frelih-Larsen, A., . . . and Sanders, M. (2011). Assessment of the potential of ecosystem-based approaches to climate change adaptation and mitigation in Europe. *Final report to the European Commission, DG Environment.* Ecologic Institute and Environmental Change Institute, Oxford University Centre for the Environment.

Nilsson, S. (2015). Global trends and possible future land use. In. Westholm, E., Lindahl, K. B., and Kraxner, F. (eds). *The Future Use of Nordic Forests.* Heidelberg, New York, Dordrecht, London: Springer International Publishing.

Nilsson, U., Fahlvik, N., Johansson, U., Lundström, A., and Rosvall, O. (2011). Simulation of the effect of intensive forest management on forest production in Sweden. *Forests*, 2(1), 373–393.

Nogués-Bravo, D., Simberloff, D., Rahbek, C., and Sanders, N. J. (2016). Rewilding is the new Pandora's box in conservation. *Current Biology*, 26(3), 87–91.

Nordblad, J. (2016). Time for politics: How a conceptual history of forests can help us politicize the long term. *European Journal of Social Theory*, 20(1), 164–182.

Nordström, E. M., Forsell, N., Lundström, A., Korosuo, A., Bergh, J., Havlík, P., . . . and Nordin, A. (2016). Impacts of global climate change mitigation scenarios on forests and harvesting in Sweden 1. *Canadian Journal of Forest Research*, 46(12), 1427–1438.

Nordström, E. M., Holmström, H., and Öhman, K. (2013). Evaluating continuous cover forestry based on the forest owner's objectives by combining scenario analysis and multiple criteria decision analysis. *Silva Fennica*, 47(4), art id. 1046.

Norgaard, R. B. (2010). Ecosystem services: From eye-opening metaphor to complexity blinder. *Ecological Economics*, 69(6), 1219–1227.

Norton, B. G. (2005). *Sustainability: A philosophy of adaptive ecosystem management*. Chicago, IL, and London: University of Chicago Press.

Nummelin, T., Widmark, C., Riala, M., Sténs, A., Nordström, E-M. and Nordin, A. (2017). Forest futures by Swedish students – developing a mind mapping method for data collection. *Scandinavian Journal of Forest Research*, *32*, Published online: 23 February 2017.

Oakerson, Ronald J. (1992). Analyzing the commons: a framework. In Bromley, D.W. (ed.). *Making the commons work: theory, practice, and policy*. San Francisco, CA: Institute for Contemporary Studies (pp. 41–62).

Öckerman, A. (1996). Hygge eller blädning? Svensk skogshistoria som miljöhistoria. In Linnér, B-J. and Svidén, J. (eds). *Miljöhistoria på väg: artiklar presenterade vid Miljöhistoriskt Möte 1995* (pp. 24–35). Linköping: Tema V Rapport 22.

Öckerman, A. (1998). Joel Wretling på Malå revir: kunskap och legitimering i det moderna skogsbruket 1945–1990. In Johansson, M. (ed.). *Miljöhistoria idag och i morgon* (pp. 239–252). Karlstad: Högskolan i Karlstad.

Ödmann, E., Bucht, E., and Nordström, M. (1982). *Vildmarken och välfärden: om naturskyddslagstiftningens tillkomst*. Lund: LiberFörlag.

OECD. (2006). *The Bioeconomy to 2030: Designing a policy agenda*. Paris: Organisation for Economic Co-operation and Development (OECD).

Olsson P., Folke C., and Hahn T. (2004). Social-ecological transformation for ecosystem management: The development of adaptive co-management of a wetland landscape IN Southern Sweden. *Ecology and Society*, 9(2).

Olwig, K. R. (1980). Historical geography and the society/nature 'problematic': the perspective of J. F. Schouw, G. P. Marsh and E. Reclus. *Journal of Historical Geography*, 6(1), 29–45.

O'Neill, B. C., Kriegler, E., Ebi, K. L., Kemp-Benedict, E., Riahi, K., Rothman, D. S., van Vuuren, D. P., Birkmann, J., Kok, K., Solecki, W., and Levy, M. (2017). The roads ahead: narratives for shared socioeconomic pathways describing world futures in the 21st century. *Global Environmental Change*, 42, 169–180.

Ong, A. (2006). *Neoliberalism as exception: mutations in citizenship and sovereignty*. Durham, NC: Duke University Press.

Oreskes, N. (2013). Why I am a presentist. *Science in Context*, 26(04), 595–609.

Östlund, L. (1993). *Exploitation and structural changes in the north Swedish boreal forest 1800–1992*. Diss. Umeå: Sveriges lantbruksuniversitet.

Östlund, L., Zackrisson, O., and Axelsson, A. L. (1997). The history and transformation of a Scandinavian boreal forest landscape since the 19th century. *Canadian Journal of Forest Research*, 27(8), 1198–1206.

Östlund, L. (ed.). (1997). *Människan och skogen: från naturskog till kulturskog?* Stockholm: Nordiska museet.

Östlund, L., Hörnberg, G., DeLuca, T. H., Liedgren, L., Wikström, P., Zackrisson, O., and Josefsson, T. (2015). Intensive land use in the Swedish mountains between AD 800 and 1200 led to deforestation and ecosystem transformation with long-lasting effects. *Ambio*, 44(6), 508–520.

Ostrom, E. (1972). Metropolitan reform: propositions derived from two traditions. *Social Science Quarterly*, 53, 474–493.

Ostrom, E. (2007). A diagnostic approach for going beyond panaceas. *Proceedings of the National Academy of Sciences*, 104(39), 15181–15187.

Ostrom, E. (2009). *Understanding institutional diversity*. Princeton, NJ: Princeton University Press.

Ostrom, E., and Cox, M. (2010). Moving beyond panaceas: a multi-tiered diagnostic approach for social-ecological analysis. *Environmental Conservation*, 37(04), 451–463.

Ostrom, E. (2011). *Governing the commons: the evolution of institutions for collective action*. Cambridge: Cambridge University Press.

Overvåg, K., Skjeggedal, T., and Sandström, C. (2015). Management of mountain areas in Norway and the persistence of local–national conflicts. *Journal of Environmental Planning and Management*, 59(7), 1186–1204.

Pahl-Wostl, C. (2002). Towards sustainability in the water sector–The importance of human actors and processes of social learning. *Aquatic Sciences*, 64(4), 394–411.

Pahl-Wostl, C. (2006). The importance of social learning in restoring the multifunctionality of rivers and floodplains. *Ecology and Society*, 11(1) 10.

Partelow, S. (2015). Coevolving Ostrom's social–ecological systems (SES) framework and sustainability science: Four key co-benefits. *Sustainability Science*,11(3), 399–410.

Pelli, P. (2008). *Review on forest sector foresight studies and exercises*. Joensuu, Finland: European Forest Institute.

Pelli, P. and den Herder, M. (2013). Foresight on future demand for forest-based products and services. *EFI Technical Report 87*. Joensuu, Finland: European Forest Institute.

Pereira, H. M., Leadley, P. W., Proença, V., Alkemade, R., Scharlemann, J. P., Fernandez-Manjarrés, J. F., . . . and Chini, L. (2010). Scenarios for global biodiversity in the 21st century. *Science*, 330(6010), 1496–1501.

Pickstone, J. V. (2001). *Ways of knowing: a new history of science, technology, and medicine*. Chicago, IL: University of Chicago Press.

Pielke Jr, R. A. (2007). *The honest broker: making sense of science in policy and politics*. Cambridge: Cambridge University Press.

Plummer, R., and Armitage, D. (2010). Integrating perspectives on adaptive capacity and environmental governance. In *Adaptive capacity and environmental governance* (pp. 1–19). Heidelberg, New York, Dordrecht, London: Springer International Publishing.

Pommerening, A., and Murphy, S. T. (2004). A review of the history, definitions and methods of continuous cover forestry with special attention to afforestation and restocking. *Forestry*, 77(1), 27–44.

Poudel, B. C., Sathre, R., Gustavsson, L., Bergh, J., Lundström, A., and Hyvönen, R. (2011). Effects of climate change on biomass production and substitution in north-central Sweden. *Biomass and Bioenergy*, 35(10), 4340–4355.

Poudel, B. C., Sathre, R., Bergh, J., Gustavsson, L., Lundström, A., and Hyvönen, R. (2012). Potential effects of intensive forestry on biomass production and total carbon balance in north-central Sweden. *Environmental Science & Policy*, 15(1), 106–124.

Power, M. (1997). *The audit society: rituals of verification*. Oxford: Oxford University Press.

Price, K., Roburn, A., and MacKinnon, A. (2009). Ecosystem-based management in the Great Bear Rainforest. *Forest Ecology and Management*, 258(4), 495–503.

Puettmann, K. J., Coates, K. D., and Messier, C. C. (2009). *A critique of silviculture: managing for complexity*. Washington, DC: Island Press.

Puettmann, K. J., Wilson, S. M., Baker, S. C., Donoso, P. J., Drössler, L., Amente, G., Harvey, B. D., Knoke, T., Lu, Y., Nocentini, S., Yoshida, T., Bauhus, J., and Putz, F. E. (2015). Silvicultural alternatives to conventional even-aged forest management-what limits global adoption? *Forest Ecosystems*, 2(1), 8.

Pülzl, H., Kleinschmit, D., and Arts, B. (2014). Bioeconomy – an emerging meta-discourse affecting forest discourses? *Scandinavian Journal of Forest Research*, 29(4), 386–393.

Radkau, J. (2012). *Wood: a history*. Cambridge: Polity Press.

Raitio, K., and Harkki, S. (2014). The disappearing chain of responsibility: Legitimacy challenges in the political governance of Finnish Forest and Park Service. *Land Use Policy*, 39, 281–291.

Ranius, T., and Roberge, J. M. (2011). Effects of intensified forestry on the landscape-scale extinction risk of dead wood dependent species. *Biodiversity and Conservation*, 20(13), 2867–2882.

Reed, M. S. (2008). Stakeholder participation for environmental management: a literature review. *Biological Conservation*, 141(10), 2417–2431.

Reed, M., Evely, A. C., Cundill, G., Fazey, I. R. A., Glass, J., Laing, A., Newig, J., Parrish, B., Prell, C., Raymond, C., and Stringer, L. (2010). What is social learning? *Ecology and Society*, 15(4).

Rist, L., Shaanker, R. U., Milner-Gulland, E. J., and Ghazoul, J. (2010). The use of traditional ecological knowledge in forest management: an example from India. *Ecology and Society*, 15(1), 3.

Rist, L., and Moen, J. (2013). Sustainability in forest management and a new role for resilience thinking. *Forest Ecology and Management*, 310, 416–427.

Rist, L., Campbell, B. M., and Frost, P. (2013). Adaptive management: where are we now? *Environmental Conservation*, 40(1), 5–18.

Rist, L., Felton, A., Nyström, M., Troell, M., Sponseller, R.A., Bengtsson, J., Österblom, H., Lindborg, R., Tidåker, P., Angeler, D.G., Milestad, R. and Moen, J. (2014). Applying resilience thinking to production ecosystems. *Ecosphere*, 5(6), 73.

Rist, L., Felton, A., Mårald, E., Samuelsson, L., Lundmark, T., and Rosvall, O. (2016). Avoiding the pitfalls of adaptive management implementation in Swedish silviculture. *Ambio*, 45(2), 140–151.

Roberge, J. M., Laudon, H., Björkman, C., Ranius, T., Sandström, C., Felton, A., . . . and Bergh, J. (2016). Socio-ecological implications of modifying rotation lengths in forestry. *Ambio*, 45(2), 109–123.

Roberge, J. M., Sténs, A., Felton, A., Johansson, J., Laudon, H., Beland-Lindahl, K., Löfmarck, E., Nilsson, U., Nordin, A., Ranius, T. Rist, L., and Widmark, C. (forthcoming) The role of ecological knowledge in conservation practice: Swedish forestry as a case.

Robin, L., Sörlin, S., and Warde, P. (eds). (2013). *The future of nature: documents of global change*. New Haven, CT, and London: Yale University Press.

Rockström, J., Steffen, W., Noone, K., Persson, Å., Chapin III, F. S., Lambin, E., . . . and Nykvist, B. (2009). Planetary boundaries: exploring the safe operating space for humanity. *Ecology and society*, 14(2).

Rodela, R. (2013). The social learning discourse: Trends, themes and interdisciplinary influences in current research. *Environmental Science and Policy*, 25, 157–166.

Rogers, E., and Weber, E. P. (2010). Thinking harder about outcomes for collaborative governance arrangements. *The American Review of Public Administration*, 40(5), 546–567.

Rotmans, J., Kemp, R., and van Asselt, M. (2001). More Evolution than Revolution. Transition Management in Public Policy. *Foresight* 3(1), 15–31.

Rotmans, J. (2005) *Societal Innovation: Between Dream and Reality Lies Complexity*, Inaugural Speech. Rotterdam: Erasmus Research Institute of Management.

Rotmans, J., and Kemp, R. (2008). Detour Ahead: A response to Shove and Walker about the perilous road of transition management. *Environment and Planning A* 40, 1006–1014.

Saarikoski, H., Tikkanen, J., and Leskinen, L. A. (2010). Public participation in practice–assessing public participation in the preparation of regional forest programs in Northern Finland. *Forest Policy and Economics*, 12(5), 349–356.

Sánchez-Azofeifa, G. A., Pfaff, A., Robalino, J.A. and Boomhower, J.P. (2007). Costa Rica's payment for environmental services program: Intention, implementation, and impact. *Conservation Biology*, 21, 1165–1173.

Sandell, K., and Sörlin, S. (2000). *Friluftshistoria: från 'härdande friluftslif' till ekoturism och miljöpedagogik: Teman i det svenska friluftslivets historia.* Stockholm: Carlssons bokförlag.

Sandström, C., and Widmark, C. (2007). Stakeholders' perceptions of consultations as tools for co-management – a case study of the forestry and reindeer herding sectors in northern Sweden. *Forest Policy and Economics*, 10(1–2), 25–35.

Sandström, C. and Lindkvist, A. (2009). Competing land use associated with Sweden's forests. External drivers affecting Swedish forests and forestry. Future Forests Working Report.

Sandström, C., Wennberg Di Gasper, S., and Öhman, K. (2013). Conflict resolution through ecosystem-based management: the case of Swedish moose management. *International Journal of the Commons*, 7(2), 549–570.

Sandström, C., and Sténs, A. (2015). Dilemmas in forest policy development–the Swedish forestry model under pressure. In *The Future Use of Nordic Forests* (pp. 145–158). Heidelberg, New York, Dordrecht, London: Springer International Publishing.

Sandström, C., Carlsson-Kanyama, A., Lindahl, K. B., Sonnek, K. M., Mossing, A., Nordin, A., . . . and Räty, R. (2016). Understanding consistencies and gaps between desired forest futures: An analysis of visions from stakeholder groups in Sweden. *Ambio*, 45(2), 100–108.

Sandström, P. (2015). *A toolbox for co-production of knowledge and improved land use dialogues.* Diss. Umeå: Sveriges lantbruksuniversitet.

Sardar, Z. (1993). Colonizing the future: the 'other' dimension of futures studies. *Futures*, 25(2), 179–187.

Sardar, Z. (2003). *Islam, postmodernism and other futures: a Ziauddin Sardar reader.* London: Pluto Press.

Sardar, Z. (2010). Welcome to postnormal times. *Futures*, 42(5), 435–444.

Sarewitz, D. (2004). How science makes environmental controversies worse. *Environmental Science & Policy*, 7(5), 385–403.

SCB. 2015. Urbanisering – från stad till land. www.scb.se/sv_/Hitta-statistik/Artiklar/Urbanisering-fran-land-till-stad/ (accessed 11 November 2016).

Scharpf, F. W. (1999). *Governing in Europe: effective and democratic?* Oxford: Oxford University Press.

Schröter, M., Zanden, E. H., Oudenhoven, A. P., Remme, R. P., Serna-Chavez, H. M., Groot, R. S., and Opdam, P. (2014, 02). Ecosystem Services as a Contested

Concept: A Synthesis of Critique and Counter-Arguments. *Conservation Letters*, 7(6), 514–523.

Schultz, L., Duit, A., and Folke, C. (2011). Participation, adaptive co-management, and management performance in the world network of biosphere reserves. *World Development*, 39(4), 662–671.

Scott, J. C. (1998). *Seeing like a state: how certain schemes to improve the human condition have failed*. New Haven, CT: Yale University Press.

SEPA (Swedish Environmental Protection Agency). (2016). Beskrivning av former för naturskydd. www.naturvardsverket.se/Miljoarbete-i-samhallet/Miljoarbete-i-Sverige/Uppdelat-efter-omrade/Naturvard/Beskrivning-former-for-naturskydd/

SFA, Swedish Forest Agency. (2014). *Swedish Statistical Yearbook of Forestry*, *www.skogsstyrelsen.se/statistics*. Jönköping.

SFS. (1962). Brottsbalk (The Criminal Code), www.riksdagen.se/sv/dokument-lagar/dokument/svensk-forfattningssamling/brottsbalk-1962700_sfs-1962–700 (accessed on 30 October 2016).

SFS. (1970). Jordabalk (The Real Property Law). www.riksdagen.se/sv/dokument-lagar/dokument/svensk-forfattningssamling/jordabalk-1970994_sfs-1970–994 (accessed on 30 October 2016).

SFS. (1971). Rennäringslagen (Swedish Reindeer Husbandry Act). www.riksdagen.se/sv/dokument-lagar/dokument/svensk-forfattningssamling/rennaringslag-1971437_sfs-1971–437 (accessed 31 October 2016).

SFS. (1998). Miljöbalken (The Environmental Code). www.riksdagen.se/sv/dokument-lagar/dokument/svensk-forfattningssamling/miljobalk-1998808_sfs-1998–808 (accessed 30 October 2016).

Siiskonen, H. (2007). The conflict between traditional and scientific forest management in 20th century Finland. *Forest Ecology and Management*, 249(1), 125–133.

Siiskonen, H. (2013). From economic to environmental sustainability: the forest management debate in 20th century Finland and Sweden. *Environment, Development and Sustainability*, 15(5), 1323–1336.

Simonsen, R., Rosvall, O., Gong, P., and Wibe, S. (2010). Profitability of measures to increase forest growth. *Forest Policy and Economics*, 12(6), 473–482.

Simonsson, P., Gustafsson, L., and Östlund, L. (2015). Retention forestry in Sweden: driving forces, debate and implementation 1968–2003. *Scandinavian Journal of Forest Research*, 30(2), 154–173.

Simonsson, P. (2016). *Conservation measures in Swedish forests: the debate, implementation and outcomes*. Diss. Umeå: Sveriges lantbruksuniversitet.

Skelcher, C. (2005). Jurisdictional integrity, polycentrism, and the design of democratic governance. *Governance*, 18(1), 89–110.

Skog för framtid. (1978). Statens offentliga utredningar 1978:6. Stockholm: Liber Förlag/Allmäna förl.

Skogsstyrelsen. (2016). Adaptiv skogsskötsel 2013–2015. Meddelanden 7 2016. Jönköping.

Skogsvårdslag. (1993). Svensk författningssamling 1979:429, reversed 1993: 553. Available online: www.svo.se/episerver4/templates/SNormalPage.aspx?id=12677.

Skyrms, B. (2014). *Evolution of the social contract*. Cambridge: Cambridge University Press.

Slocombe, D. S. (1993). Implementing ecosystem-based management. *BioScience*, 43(9), 612–622.

Smith, A., and Kern, F. (2009). The transitions storyline in Dutch environmental policy. *Environmental Politics*, 18(1), 78–98.

Söderqvist, T. (1986). *The ecologist: from merry naturalists to saviours of the nation: a sociologically informed narrative survey of the ecologization of Sweden 1895–1975*. Stockholm: Almqvist & Wiksell.

Sölvell, O., Zander, I., and Porter, M. E. (1992). *Advantage Sweden*. Basingstoke, UK: Macmillan.

Sondeijker, S. (2009). Imagining Sustainability: methodological building blocks for Transition Scenarios. The Netherlands: DRIFT, Erasmus University Rotterdam.

Sörlin, S. (1988). *Framtidslandet: debatten om Norrland och naturresurserna under det industriella genombrottet*. Stockholm: Carlssons.

Sörlin, S. (2011). The contemporaneity of environmental history: negotiating scholarship, useful history, and the new human condition. *Journal of Contemporary History*, 46(3), 610–630.

Soyez, D. (2000). Anchored locally–linked globally. Transnational social movement organizations in a (seemingly) borderless world. *GeoJournal*, 52(1), 7–16.

Staffas, L., Gustavsson, M., and McCormick, K. (2013). Strategies and policies for the bioeconomy and bio-based economy: An analysis of official national approaches. *Sustainability*, 5(6), 2751–2769.

Sténs, A., and Sandström, C. (2013). Divergent interests and ideas around property rights: the case of berry harvesting in Sweden. *Forest Policy and Economics*, 33, 56–62.

Sténs, A., and Sandström, C. (2014). Allemansrätten in Sweden: a resistant custom. *Landscapes*, 15(2), 106–118.

Sténs, A., Bjärstig, T., Nordström, E. M., Sandström, C., Fries, C., and Johansson, J. (2016). In the eye of the stakeholder: the challenges of governing social forest values. *Ambio*, 45(2), 87–99.

Stilgoe, J., Owen, R., and Macnaghten, P. (2013). Developing a framework for responsible innovation. *Research Policy*, 42(9), 1568–1580.

Stirling, A. (2008). 'Opening up' and 'closing down' power, participation, and pluralism in the social appraisal of technology. *Science, Technology & Human values*, 33(2), 262–294.

Stirling, A. (2010). Keep it complex. *Nature*, 468(7327), 1029–1031.

Stjernquist, P. (1973). *Laws in the forests: a study of public direction of Swedish private forestry*. Lund, Sweden: Gleerup.

Streyffert, T. (1961). Skogsforskningens mål och medel. In Lindh, E. (ed.). *Skogen och skogsbruket* (pp. 412–428). Stockholm: Svensk Litteratur.

Surel, Y. (2000). The role of cognitive and normative frames in policy-making. *Journal of European Public Policy*, 7(4), 495–512.

Swedish Environmental Objectives Council. (2006). *Sweden's environmental objectives: buying into a better future*. Stockholm: Swedish Environmental Protection Agency, www.naturvardsverket.se/Documents/publikationer/620–1251–7.pdf

Takacs, D. (1996). *The idea of biodiversity: Philosophies of paradise*. Baltimore, MD: Johns Hopkins University Press.

TEEB, and Kumar, P (ed.). (2010). *The economics of ecosystems and biodiversity ecological and economic foundations*. London and Washington: Earthscan.

Teutschbein, C., Grabs, T., Karlsen, R. H., Laudon, H., and Bishop, K. (2015). Hydrological response to changing climate conditions: Spatial streamflow variability in the boreal region. *Water Resources Research*, 51(12), 9425–9446.

Tham, Å. (1994). Crop plans and yield predictions for Norway spruce (Picea abies (L.) Karst.) and birch (Betula pendula Roth & Betula pubescens Ehrh.) mixtures (No. 195).

The International Property Right Index. (2016). http://internationalpropertyrights index.org/ (accessed 17 February 2017).

Tomppo, E., Gschwantner, T., Lawrence, M., McRoberts, R. E., Gabler, K., Schadauer, K., . . . and Cienciala, E. (2010). *National forest inventories: Pathways for Common Reporting*. Heidelberg, New York, Dordrecht, London: Springer Science + Business Media.

Toppinen, A., and Kuuluvainen, J. (2010). Forest sector modelling in Europe – the state of the art and future research directions. *Forest Policy and Economics*, *12*(1), 2–8.

Torfing, J. (2012). *Interactive governance: advancing the paradigm*. Oxford: Oxford University Press.

Tunlid, A. (2007). Ett konfliktfyllt fält: förtroende och trovärdighet inom miljöforskningen. In Agrell, W. (ed.). *Forskningens gråzoner: tillrättalägganden, anpassning och marknadsföring i kunskapsproduktion* (pp. 134–151). Stockholm: Carlsson.

Turnhout, E., Waterton, C., Neves, K., and Buizer, M. (2013). Rethinking biodiversity: from goods and services to 'living with'. *Conservation Letters*, *6*(3), 154–161.

Uggla, Y., and Lidskog, R. (2016). Climate risks and forest practices: Forest owners' acceptance of advice concerning climate change. *Scandinavian Journal of Forest Research*, *31*(6), 618–625. doi:10.1080/02827581.2015.1134648

Uggla, Y., Forsberg, M., and Larsson, S. (2016). Dissimilar framings of forest biodiversity preservation: Uncertainty and legal ambiguity as contributing factors. *Forest Policy and Economics*, *62*, 36–42.

Underdånigt betänkande och förslag angående åtgärder för befrämjande af en förbättrad skogshushållning. (1856). Stockholm: 1856 års skogshushållnings-kommitté.

UNFCCC. (2015). The Paris Agreement. http://unfccc.int/paris_agreement/items/ 9485.php (Accessed on 17 February 2017).

United Nations Economic Commission for Europe. (2011). *The European Forest Sector Outlook Study II, 2010–2030* (No. 28). Geneva: United Nations Publications.

Van der Plas, F., Manning, P., Allan, E., Scherer-Lorenzen, M., Verheyen, K., Wirth, C., . . . and Barbaro, L. (2016). Jack-of-all-trades effects drive biodiversity-ecosystem multifunctionality relationships in European forests. *Nature Communications*, *7*, 11109.

Van Huijstee, M. M., Francken, M., and Leroy, P. (2007). Partnerships for sustainable development: a review of current literature. *Environmental Sciences*, *4*(2), 75–89.

Van Notten, P. W., Rotmans, J., Van Asselt, M. B., and Rothman, D. S. (2003). An updated scenario typology. *Futures*, *35*(5), 423–443.

Veenman, S., Liefferink, D., and Arts, B. (2009). A short history of Dutch forest policy: The 'de-institutionalisation' of a policy arrangement. *Forest Policy and Economics*, *11*(3), 202–208.

Voß, J.P., and Kemp, R. (2005). Reflexive Governance for Sustainable Development – Incorporating feedback in social problem solving. Paper for ESEE Conference 14–17 June, 2005, in Lisbon. Special session on transition management.

Voß, J-P., Kemp, R. and Bauknecht, D. (2005). Reflexive Governance: A View on the Emerging Path, in J-P. Voss, D. Bauknecht and Kemp R. (eds). *Reflexive Governance for Sustainable Development*, Edward Elgar, Cheltenham.

Voß, J. P., Newig, J., Kastens, B., Monstadt, J., and Nölting, B. (2007). Steering for sustainable development: A typology of problems and strategies with respect to ambivalence, uncertainty and distributed power. *Journal of Environmental Policy & Planning*, 9(3–4), 193–212.

Voß, J. P., and Bornemann, B. (2011). The politics of reflexive governance: challenges for designing adaptive management and transition management. *Ecology and Society*, 16(2).

VROM (Ministy of Housing, Spatial Planning and the Environment). (2001). *Where there is a will there is a world: working towards sustainability. Fourth national environmental policy plan*. The Netherlands: VROM.

Walker, B., C. S. Holling, S. R. Carpenter, and A. Kinzig. (2004). Resilience, adaptability and transformability in social–ecological systems. *Ecology and Society*, 9(2), 5, www.ecologyandsociety.org/vol9/iss2/art5/

Walker, B., Gunderson, L., Kinzig, A., Folke, C., Carpenter, S., and Schultz, L. (2006). A handful of heuristics and some propositions for understanding resilience in social-ecological systems. *Ecology and Society*, 11(1), 13.

Walker, B., Sayer, J., Andrew, N.L. and Campbell, B. (2010). Should Enhanced Resilience be an Objective of Natural Resource Management Research for Developing Countries? *CROP Science*, 50: S-10-S-19.

Walker, B., and Salt, D. (2012). *Resilience thinking: sustaining ecosystems and people in a changing world*. Washington, DC: Island Press.

Wall, D. H., Bardgett, R. D., Covich, A. P., and Snelgrove P. V. R. (2004). The need for understanding how biodiversity and ecosystem functioning affect ecosystem services in soils and sediments. In Wall, D. H. (ed.). *Sustaining Biodiversity and Ecosystem Services in Soils and Eediments* (pp. 1–12). Washington, DC: Island Press.

Wallace, A. R. (1863). On the physical geography of the Malay Archipelago. *Journal of the Royal Geographical Society of London*, 33, 217–234.

Wallin, I. (2017). *Forest management and governance in Sweden: a phronetic analysis of social practices*. Diss. Alnarp: Sveriges lantbruksuniversitet.

Wallmo, U. (1897). *Rationell skogsafverkning: praktiska råd till såväl större som mindre enskilde skogsägare samt svar på en fråga för dagen*. Stockholm: Fritzen in Komm.

Walters, C. J., and Hilborn, R. (1978). Ecological optimization and adaptive management. *Annual Review of Ecology and Systematics*, 9(1), 157–188.

Walters, C. J., and Holling, C. S. (1990). Large-scale management experiments and learning by doing. *Ecology*, 71(6), 2060–2068.

Walters, C. J. (2007). Is adaptive management helping to solve fisheries problems?. *Ambio*, 36(4), 304–307.

Warde, P. (2006). *Ecology, economy and state formation in early modern Germany*. New York: Cambridge University Press.

Warde, P. (2011). The invention of sustainability. *Modern Intellectual History*, 8(01), 153–170.

Warde, P., and Sörlin, S. (2015). Expertise for the future: the emergence of environmental prediction c. 1920–1970. In Andersson, J. and Rindzevičiūtė, E. (eds). *The struggle for the long-term in transnational science and politics: forging the future* (pp. 38–62). London & New York: Routledge.

Waylen, K. A., Hastings, E. J., Banks, E. A., Holstead, K. L., Irvine, R. J., and Blackstock, K. L. (2014). The need to disentangle key concepts from Ecosystem Approach jargon. *Conservation Biology, 28*(5), 1215–1224.

Webler, T., Kastenholz, H., and Renn, O. (1995). Public participation in impact assessment: a social learning perspective. *Environmental Impact Assessment Review, 15*(5), 443–463.

Wellstead, A., Howlett, M., and Rayner, J. (2013). The neglect of governance in forest sector vulnerability assessments: Structural-functionalism and 'black box' problems in climate change adaptation planning. *Ecology and Society, 18*(3).

Wennberg DiGasper, S. (2008). *Natural resource management in an institutional disorder: the development of adaptive co-management systems of moose in Sweden.* Diss. Luleå: Luleå University of Technology.

Westholm, E., Lindahl, K. B., and Kraxner, F. (eds). (2015). *The Future Use of Nordic Forests.* Heidelberg, New York, Dordrecht, London: Springer International Publishing.

Widman, U. (2015). Shared responsibility for forest protection? *Forest Policy and Economics, 50,* 220–227.

Widman, U. (2016). Exploring the role of public–private partnerships in forest protection. *Sustainability, 8*(5), 496.

Widmark, C. (2009). Forestry and reindeer husbandry in northern Sweden–the development of a land use conflict. *Rangifer, 26*(2), 43–54.

Williams, M. (2003). *Deforesting the earth: From prehistory to global crisis.* Chicago, IL: University of Chicago Press.

Winkel, G., and Sotirov, M. (2011). An obituary for national forest programmes? Analyzing and learning from the strategic use of 'new modes of governance' in Germany and Bulgaria. *Forest Policy and Economics, 13*(2), 143–154.

Winkel, G., Gleißner, J., Pistorius, T., Sotirov, M., and Storch, S. (2011). The sustainably managed forest heats up: discursive struggles over forest management and climate change in Germany. *Critical Policy Studies, 5*(4), 361–390.

Winkel, G. (2014). When the pendulum doesn't find its center: Environmental narratives, strategies, and forest policy change in the US Pacific Northwest. *Global Environmental Change, 27,* 84–95.

Winkel, G., and Sotirov, M. (2014). Whose integration is this? European forest policy between the gospel of coordination, institutional competition, and new spirits of integration. *Environment and Planning C: Government and Policy.* doi: 10.1068/c 1356j.

Wirén, Erik. (1985). *Allemans skog: slutrapport för projektet: Samhället och skogen.* 1. uppl. Stockholm: LiberFörlag.

Worster, D. (1994). *Nature's economy: A history of ecological ideas.* Cambridge: Cambridge University Press.

Wunder, S., Engel, S., and Pagiola, S. (2008). Taking Stock: A Comparative Analysis of Payments for Environmental Services Programs in Developed and Developing Countries. *Ecological Economics 65*(4), 834–852.

Yaffee, S. L. (1999). Three faces of ecosystem management. *Conservation Biology, 13*(4), 713–725.

Zachrisson, A. (2009). *Commons protected for or from the people?: Co-management in the Swedish mountain region?* Diss. Umeå: Umeå universitet.

Zaremba, M. (2012). *Skogen vi ärvde.* Stockholm: Weyler.

Index